Analytical Reaction Gas Chromatography

by Viktor G. Berezkin

A. V. Topchiev Institute of Petrochemical Synthesis
Academy of Sciences of the USSR, Moscow

Translated from Russian

Translation Editor: L. S. Ettre

Chief Applications Chemist, Analytical Division
The Perkin-Elmer Corporation, Norwalk, Connecticut

℗ PLENUM PRESS • NEW YORK • 1968

Born in Moscow in 1931, Viktor Grigor'evich Berezkin was graduated from the Chemistry Department of the M. V. Lomonosov Moscow State University in 1954, and earned his candidate's degree there in 1962. The author of 80 published works in the field of gas chromatography and radiation chemistry, Berezkin is at the present time the head of the laboratory at the A. V. Topchiev Institute of Petrochemical Synthesis of the Academy of Sciences of the USSR.

The original Russian text, published by Nauka Press in Moscow in 1966 for the A. V. Topchiev Institute of Petrochemical Synthesis of the Academy of Sciences of the USSR, has been corrected by the author for this edition.

Виктор Григорьевич Берёзкин

Аналитическая реакционная газовая хроматография

ANALITICHESKAYA REAKTSIONNAYA GAZOVAYA KHROMATOGRAFIYA

ISBN-13: 978-1-4684-0711-2   e-ISBN-13: 978-1-4684-0709-9
DOI: 10.1007/978-1-4684-0709-9

Library of Congress Catalog Card Number 68-21473

ANALYTICAL REACTION
GAS CHROMATOGRAPHY

Foreword to the English Edition

One of the main tendencies in the development of contemporary analytical chemistry is toward the wide use of physical methods for the analysis of different compounds and their mixtures. However, the most effective use of these methods for analytical purposes can be achieved only in combination with chemical methods. This fact is also valid for gas chromatography.

In recent years a new prospective direction has been formulated in gas chromatography – analytical gas chromatography with chemical reactions, in which (for analytical purposes) the compounds being analyzed are subjected to chemical reactions in conjunction with the chromatographic separation.

At the present time this method is developed to a higher degree than the standard methods and techniques of gas chromatography, and it is fair to assume that in the near future a gas chromatograph incorporating standard reactors will become as common as the chromatograph incorporating a number of standard detectors is now.

In this book an attempt has been made to indicate the methods and the paths for the development of analytical gas chromatography with reactions – or analytical reaction gas chromatography. It is hoped that the book will facilitate the development of interest in further studies in this field.

The translation of this book will, obviously, also be useful because it helps the reader to become familiar with the literature of this field published in the USSR. There have been some changes and additions to the text of the original Russian book for the English edition, and a new chapter on selective chemically active sorbents has been added.

In conclusion the author expresses his gratitude to the publishers for their attention to his work.

Moscow   V. G. Berezkin
August, 1967

Foreword

The most universal and effective method for the analysis of complex multicomponent mixtures of volatile substances is gas chromatography. However, there are a number of limitations associated with the classical variation of this technique which retard its development and the further expansion of its application:

1) the identification of the components of a complex mixture of unknown composition is in itself a complex and difficult problem, unless the homologous series of the component to be identified is known;

2) the overlapping of chromatographic peaks for several compounds makes it difficult, and in a number of cases impossible, to carry out qualitative and quantitative analysis of these components, and leads to the necessity of using several columns of different polarities or to the use of columns with very high efficiency;

3) the direct analysis of unstable and nonvolatile compounds is impossible;

4) the difficulty of quantitative chromatographic analysis using thermal conductivity detectors increases with the necessity of determining individual response (calibration) factors; the insensitivity of the flame ionization detector to a number of substances (inorganic gases) leads to the necessity of introducing additional operations (preliminary concentration of trace components) in connection with thermal conductivity detectors.

The directed use of chemical conversion of the compounds analyzed usually makes it possible to remove the limitations cited above. The combination of two analytical methods (chromatographic and chemical) led to the creation of a new field of gas chromatography–analytical gas chromatography with reactions, or analytical reaction gas chromatography.

At the present time the methods of this reaction gas chromatography are being used successfully and rather widely in gas chromatographic practice. The total number of scientific publications on individual questions of reaction gas chromatography exceeds 800.

However, in spite of a large number of original studies in the literature devoted to individual problems, there are no reports which study in detail all the questions associated with reaction gas chromatography.

In this book the main results in the theory, methods, and application areas of reaction gas chromatography are systematized and generalized; further, an attempt is made to point out some prospective developments along these lines. The goal of the author is to draw the attention of analytical and physical chemists to this interesting and prospective region of chromatographic analysis.

Chapter V "Analysis of Impurities" was written with the participation of O. L. Gorshunov, Chapter VII, "Elemental Analysis," was written in conjunction with V. M. Fateeva.

The author expresses his gratitude to A. L. Rozental and M. I. Yanovskii, who read Chapter II with a number of critical comments.

CONTENTS

Introduction

In the last decade a number of new directions have been vigorously developed in gas chromatography: analysis with temperature programming, capillary chromatography, preparative chromatography, stage chromatography, etc. In comparison with the classical variations of gas–liquid partition chromatography developed by James and Martin [1], all of these directions are characterized by the new use of some physical-variable parameter such as temperature, column diameter, etc.

The use of chemical conversions of the components being analyzed (i.e., the introduction of a new "chemical" variable) lead to the formulation of a new direction in gas chromatography – gas chromatography with reactions, or reaction gas chromatography.

Apparently, the first analytical studies using chemical transformations in the chromatographic system were carried out in 1955 by Ray [2], Green [3], and Martin and Smart [4]. The term "reaction gas chromatography" was suggested by Drawert [5].

Reaction gas chromatography is a variation of gas chromatography in which chemical reaction in a single chromatographic system [6] is used for analytical purposes. As with any general definition, additions and comments are required:

1. Since gas chromatography with chemical reactions is a variation of gas chromatography, it is necessary to give first a general definition of gas chromatography. According to Keulemans [7], "Chromatography is a physical method of separation in which the components to be separated are distributed between two phases,

1

one of the phases constituting a stationary bed of large surface area, the other being a fluid that percolates through or along the stationary bed." According to this definition, in gas chromatography, the moving phase is a gas flow under the action of which the components of the mixture being analyzed are moved in the column along a layer of the stationary phase at different rates, determined by their sorbability. A nonselective detector, placed after the chromatographic column, registers the separation of the components into individual chromatographic zones.

At the present time this definition is not completely accurate with respect to the development of the methods of reaction chromatography. Chromatography should be considered as a physicochemical method of separation, in which the distribution constants are mainly determined by chemical forces or by the formation of complex compounds of the substances being chromatographed in the stationary phase. As examples for complex-forming chemical-chromatographic separation processes the separation of unsaturated hydrocarbons on a stationary phase consisting of the solution of a silver salt in a polar solvent [8] or the separation of amines on a zinc or copper salt of stearic acid [9] should be mentioned.

2. It should be noted that analysis is not the only goal in reaction gas chromatography. Reaction gas chromatography is successfully used to study the kinetics of catalytic processes and for the rapid measurement of such physicochemical values as the rate constants of chemical reactions, activation energies, etc. [10, 11]. However, in this book we restrict ourselves to the analytical use of reaction gas chromatography.

3. In analytical reaction gas chromatography, chemical reactions can be used at all stages of the chromatographic analysis — from the introduction of the sample to its detection.

The main analytical goal in using chemical reactions is to simplify the solution of analytical problems and expand the useful area of gas chromatography. It is particularly advantageous to apply reaction gas chromatography in those cases where the direct use of general gas-chromatographic methods is impossible or can be done only with significant difficulties (analysis of polymers, elemental analysis, etc.).

In reaction gas chromatography, the efficiency of the chromatographic column and the sensitivity of the detectors, as a rule, remain unchanged; however, as the result of the chemical transformation of the sample mixture, new compounds are formed which, in the general case, result in a change in the partition coefficients and in the sensitivity of the detectors, the basic parameters of chromatographic analysis.

4. Chemical conversions in analytical reaction gas chromatography always take place in an integral chromatographic system, in reactors or in the column itself. Methods in which the sample is first subjected to a purposeful conversion, independent of the chromatographic apparatus, and only then is analyzed by gas chromatography, should not be considered within the framework of analytical reaction gas chromatography.

As a conclusion, we may state that as the result of the organic blending of gas-chromatographic and chemical analytical methods a new variation of gas chromatography – analytical reaction gas chromatographic – was developed. The combined use of the two methods (chromatographic and chemical) resulted in the formation of a qualitative, new analytical method, which has much broader possibilities than either of the two individual methods. This is understandable, since in reaction gas chromatography the information obtained is based simultaneously on both the chromatographic and chemical properties of the compounds being analyzed.

Reaction gas chromatography is characterized by special, specific techniques and apparatus. New, general analytical methods have been developed, such as the method of subtraction, methods of elemental analysis, the kinetic method, etc., which represent considerable progress in analysis of complex mixtures. As a rule, a new element – the reactor – appears in the apparatus and its effect on the chromatographic parameters must usually be taken into consideration.

Reaction gas chromatography is a new, rapidly developing branch of gas chromatography. In connection with the already great practical value of this technique, it should be expected that the circle of its application will be continuously expanded.

LITERATURE CITED

1.   A. T. James and A. J. P. Martin, Biochem. J., 50:679 (1952).
2.   N. H. Ray, Analyst, 80:853, 957 (1955).

3.   G. E. Green, Nature, 180:295 (1957).
4.   A. E. Martin and J. Smart, Nature, 175:422 (1955).
5.   F. Drawert, in "Gas Chromatography in 1961," Moscow,
     Gostoptekhizdat, 1963, p. 10.
6.   V. G. Berezkin and O. L. Gorshunov, Usp. Khim., 34:1108
     (1965).
7.   A. I. M. Keulemans, Gas Chromatography, Reinhold, New
     York, 1957 [Russian translation: IL, Moscow, 1959].
8.   M. E. Bednas and D. S. Russel, Can. J. Chem., 36:1272 (1958).
9.   C. S. G. Phillips, Gas Chromatography, Butterworths, Lon-
     don, 1956, p. 57.
10. R. J. Kokes, H. Tobin, and P. H. Emmett, J. Am. Chem. Soc.,
     77:5860 (1955).
11. S. Z. Roginskii, M. I. Yanovskii, and G. A. Gasiev, Dokl.
     Akad. Nauk SSSR, 140:1125 (1961).

Chapter I

Main Directions in the Development of Analytical Reaction Gas Chromatography

The main goal in using chemical reactions in gas chromatography is the solution of definite analytical problems which are difficult to solve within the framework of the classical variations of gas chromatography [2].

As a result of using chemical conversions a chemical transformation of the original sample mixture takes place which generally results in a different separation of the component of the new mixture and in a change in the sensitivity of the detectors.

If for example the mixture being analyzed consists of a number of compounds, some of which cannot be separated, it is possible to determine the composition of this mixture on the basis of the simultaneous use of chromatographic separation and chemical reactions.

The problem involved here is to obtain, as the result of a chemical reaction, a new mixture the components of which can be separated (or detected) better than in the case of the original sample mixture. In this case the values of retention times and sensitivities of detection can change over broad limits. For purposes of classification it is useful to consider the following three cases: 1) $t_n \gg t_i$, 2) $t_n \ll t_i$, 3) $t_n \approx t_i$, where t_n is the retention time of the new compound and t_i is the retention time of the initial compound. Usually the first and second case, when the values of the retention change most sharply, are of the greatest interest. In the first

5

case the substance is not eluted from the reactor to any great extent and consequently the reactor can be placed either before or after the column [3]. In the second case the elution time decreases sharply and the reactor must be placed before the chromatographic column (or between two columns).

The directed change in sensitivity of the detectors to the compounds being analyzed is also very important in solving the task which has been set up. Thus for example one and the same result can be obtained either by the conversion of some of the sample components into undetectable or nonvolatile compounds. In a number of cases (particularly in the analysis of inorganic gases by means of flame ionization detectors) it is important to convert the undetectable inorganic compounds into detectable organic compounds. In connection with this it is convenient to separate the following typical cases of changes in sensitivity: 1) $E_n \ll E_i$, 2) $E_n \gg E_i$, 3) $E_n \approx E_i$, where E_n is the sensitivity for the compound which is formed and E_i is the sensitivity for the initial compound.

The main directions in the development of analytical reaction gas chromatography will be discussed in detail in this book.

NEW APPLICATIONS
OF GAS CHROMATOGRAPHY

The classical variation of gas chromatography provides for the analysis of compounds which are thoroughly stable at a given temperature and which have a sufficient high vapor pressure (usually greater than 10 mm). Therefore it is not possible to analyze directly a number of compounds which are of practical importance (unstable peroxides, polymers, carbonates, etc.) by gas chromatographic methods. Great difficulties are also encountered in analyzing reactive polar compounds (sulfur dioxide, hydrogen bromide, hydrogen chloride, etc.).

Nonvolatile and unstable compounds may be characterized by the gas chromatographic "spectrum" of the stable and volatile products formed during their breakdown. Thus for example the gas chromatographic method for analyzing polymeric substances by means of the chromatogram of the volatile products from their pyrolysis [4, 5] has gained wide use. This method makes it possible to identify the materials being pyrolyzed, to determine the composition of a copolymer and also in some cases the structures of the polymer. As an example the work of Voigt [6] can be cited, in which it was shown that the chromatograms of the pyrolysis products of various polyolefins (polyethylene, polypropylene, poly-butene-1, poly-4-methylpentene-1) are quite different, and methods have been developed for the quantitative determination of the monomer concentrations in copolymers of ethylene and propylene. The pyrolysis method was also satisfactorily applied to the analysis of solid substances, e.g., barbiturates and alkaloids [7], and to the qualitative identification of volatile compounds [8]. The particular value of the latter method is apparently the possibility of identification of isomeric hydrocarbons.

Pyrolysis is an important, but not the only reaction which makes it possible to expand the field of analyzed compounds. The sensitive method for analysis of carbonates [9] can be cited as another interesting example. This method is based on the separation of carbon dioxide from the carbonates by the action of hydrochloric acid solution and on the subsequent analysis of aliquot parts of the gas selected from the reaction vessel which is connected in a single system with the gas chromatograph.

A further example for the expansion of the application of gas chromatography is the analysis of reactive sulfur trioxide in a mixture with sulfur dioxide [10]. In a preliminary column the sulfur trioxide is quantitatively reacted with oxalic acid, giving off an equivalent amount of carbon dioxide and carbon monoxide. The resulting gas mixture is then separated on chromatographic columns prepared with phenyl cellosolve and silica gel.

IMPROVEMENT IN THE SEPARATION AND
IDENTIFICATION OF SAMPLE COMPONENTS

It is difficult to separate these two main problems in reaction gas chromatography because the chemical reaction, which leads to the formation of a new more easily separated mixture, is, as a rule, simultaneously the means of identification, since the reactivity of the substances is directly related to their structure.

Improvement in separation (or simplification of the problem being solved; for example reduction of the temperature of analysis, or decreasing the analysis time) can be achieved by the conversion of the original sample mixture into a new mixture with the same (or greater) number of components, or by simplifying the mixture, usually by forming nonvolatile compounds.

Thus, for example, the gas chromatographic analysis of a mixture of organic substances with water represents a serious problem due to the asymmetric water peak and the similar retention times for water and some organic oxygen-containing compounds; this problem is particularly evident when small concentrations of water in organic compounds are to be determined. To overcome these difficulties, special methods were developed [11, 12] for the determination of traces of water in organic solutions and for the analysis of aqueous solutions. These methods are based on the quantitative conversion of water vapors into acetylene through reaction with calcium carbide. The formed acetylene is rapidly eluted as a narrow peak. This method had been applied for the analysis of mixtures of water with hydrocarbons, aldehydes, ketones, esters, and alcohols.

Drawert [13] suggested using a complex multistage reactor in which the original compounds are converted into propylene and ethane for the rapid determination of alcohol and glycerine in aqueous solution. The gas chromatographic determination of the formed products (propylene and ethane) needs a simpler apparatus and lower temperatures than the determination of the original substances (glycerine and ethanol).

The chromatographic analysis of complex polar mixtures is usually a difficult problem, particularly because of the asymmetrical peaks which are often obtained for the polar substances. Therefore, in a number of cases it may be convenient to convert the polar components of the sample into nonpolar substances having known retention times. Thus Drawert [14] recommends that polar alcohols in aqueous solutions be determined in the form of the corresponding nonpolar products (such as olefins, paraffins, nitrous esters) into which the alcohols are converted in a special reactor placed in front of the chromatographic column.

The hydrogenation of unsaturated compounds and the dehydrogenation of cyclohexane homologs result in the formation of saturated or aromatic hydrocarbons, which have been chromatographically studied in greatest detail. These conversions result in an oriented change (as compared to the analysis of the original sample) in the retention times of the peaks for the reacting compounds which can be used for the identification of the substances analyzed [15, 16]. Unfortunately, the use of hydrogenation does not always permit the exact identification of the reacting component (for example the determination of the position of the double bond in an olefin) although it usually permits the rapid determination of the carbon structure of the substances. If a complete qualitative analysis is to be carried out a series of different chemical reactions have to be applied in combination with different gas chromatographic identification methods [17, 18].

Reactions which simplify the mixture to be analyzed are widely used in chromatographic analysis. In the subtraction method two chromatographic analyses of the original sample mixture are carried out: first, a conventional analysis without any chemical reaction, and second, an analysis using a chemical adsorbent which forms nonvolatile compounds with certain components of the original sample. The peaks of the reacting components are absent in the chromatogram from the second analysis. The subtraction method also permits the determination of the concentrations of components which cannot be separated chromatographically if one of them reacts with a selective adsorbent. Simultaneously this method makes it possible to carry out group identification of the reacting compounds. Martin [3] suggested the application of the subtraction method for the quantitative determination of unsaturated components of hydrocarbon mixtures. He used a small reactor filled with silica gel treated with concentrated sulfuric acid in order to retain the unsaturated compounds. It was shown that monoolefins, diolefins, cycloolefins, and acetylene homologues (up to 8 carbon atoms) are retarded from carrier gas flow at 20-50°C.

Hoff and Feit [19] suggested carrying out subtraction reactions in a syringe. Gaseous samples (2-5 ml) were introduced into the usual injection syringe, the walls of which were wet with a selective agent which formed practically nonvolatile compounds with the sample component. Concentrated sulfuric acid reacted with all the substances except saturated hydrocarbons and benzene; metallic sodium removed all the substances except hydrocarbons and ethers; finally, reactions with gaseous ozone resulted in the removal of unsaturated compounds from the sample mixture.

The rate of the chemical reaction depends greatly on the structure of the reacting compounds which can be used to identify the compounds being analyzed chromatographically. Thus Gil-Av and Herzberg-Minzly [20] suggested the method of "partial subtraction chromatography" in which the data on the retention volumes (chromatographic characteristics) and on the rates of the chemical reaction (kinetic characteristics) were used simultaneously to identify the compounds. In this work a column-reactor containing chloromaleic anhydride was used to identify diene isomers. The identification was based on the different reactivity of the isomers: trans-isomers react readily with chloromaleic anhydride while cis-isomers react much more slowly or do not react at all. Therefore upon increasing the duration of the analysis (reaction), the relative area for the trans-isomer will decrease. The trans- and cis-isomers of 1,3-pentadiene, 2,4-hexadiene, and others dienic compounds were identified in this way.

THE USE OF CHEMICAL REACTIONS TO DETECT THE COMPOUNDS BEING ANALYZED

The final result of chromatographic analysis depends not only on the efficiency and selectivity of the chromatographic separation but also on the characteristics of the type of detector used.

Franc and Jokl [21] developed a method for the analysis of aromatic hydrocarbons, among them m- and p-xylenes, which are not separated on packed columns prepared with the usual phases. In order to determine the concentration of the isomers from the overlapping peak, each sample was analyzed twice by using an optical ultraviolet absorption detector at two different wavelengths.

The use of chemical reactions makes it possible to change the observed sensitivity of any detector by converting the sample components into detectable and nondetectable compounds before they are put into the detecting device. Such a method permits not only the analyzing of inseparable components by carrying out two analyses with the different sensitivity of the detector with respect to the inseparable components, but also the identification of them. In the latter case, chemical reactions fulfill the same function as a qualitative identification reaction. An example of this method is the analysis of complex mixtures containing nitrogen-substituted substances [22]. There, if the individual components after separation on the column are oxidized, the products will be carbon dioxide, water, and nitrogen dioxide. The nitrogen oxides are reduced to nitrogen and carbon dioxide, and water is selectively adsorbed on the 5-Å molecular sieve so that only nitrogen reaches the thermal conductivity detector, and its amount will be indicative of the original nitrogen-containing organic substance. The whole process takes place in a continuous system with helium carrier gas; the oxidation is performed at 800°C by copper oxide and the reduction with finely divided copper, also at 800°C.

Another important direction in the application of chemical reactions in the detection is the use of qualitative reactions to determine the nature of the compound eluted from the chromatographic column. Various techniques are used to accomplish these reactions: they are carried out in removable test tubes (fraction collectors) through which the carrier gas passes [23], or in a thin layer which is constantly moving with respect to the effluent gas from the chromatograph at the velocity of the recorder chart paper [24]; another method is to identify halogen-containing compounds with the help of the Beilstein reaction by observing the change in the color of a flame [25]. In our opinion all of these methods can be considered as the introduction of an additional chemical detector into the chromatographic system.

An interesting example for a detector sensitive to oxidizing agents was suggested by Gudzinowicz and Smith [26]. This new detector is based on the fact that strong oxidizing agents leaving the column will react with a radioactive nonvolatile compound; as a result of this reaction, a volatile radioactive compound will be liberated which can be recorded by a Geiger counter.

This method was used to determine very small concentrations of an oxidizing agent (fluorine, chlorine, bromine, nitrosylchloride, nitrogen dioxide, etc.) in gases, the nonvolatile radioactive compound was ^{85}Kr-quinol clathrate, and during the reaction ^{85}Kr atoms were liberated.

The flame ionization detector is characterized by high sensitivity with respect to organic compounds, but is almost insensitive to a number of simple substances such as nitrogen, oxygen, argon, carbon monoxide, carbon dioxide, carbon sulfoxide, etc. Chemical reactions can be used to convert the undetectable compounds into detectable compounds which, in turn, can be detected with help of the flame ionization detector.

In a number of studies [27-30] it was shown that it is possible to achieve a sensitive detection of some of the inorganic compounds listed above by their quantitative conversion into methane, for which the flame ionization detector has a very high sensitivity. These methods are of great practical interest since such inorganic impurities (e.g., the oxides of carbon, carbon sulfoxide, etc.) greatly influence the quality of monomers and other high-purity products.

The use of chemical reactions in a number of cases is also justified when using thermal conductivity detection [31]. By converting the organic substances to be analyzed into carbon dioxide or hydrogen it is possible to increase the sensitivity of detection, to simplify the quantitative calculation, and to use a simpler apparatus (e.g., the separation can be carried out at high temperatures and the detection at room temperature). For these reasons, conversion is widely used in the practice of gas chromatography.

The examples cited above illustrate the main directions in the development of analytical reaction gas chromatography. The methods of this technique are widely used in different areas of gas chromatography: in the analysis of complex mixtures, in the identification of unknown components, or in the improved detection of certain substances. The result is the expansion of the application areas of gas chromatography. In our opinion, the further development of analytical reaction gas chromatography will take place by means of exploring new methods for using chemical reactions in gas chromatographic analysis (the characteristic feature being the use of several different chemical conversions in a single analysis), as well as by using new reactions in known methods. An important task of analytical reaction gas chromatography is also the development of methods which are directed toward the suppression of secondary undesired chemical reactions and of the adsorption of the compounds being analyzed in the column (for example, see [32]).

The greatest progress in analytical reaction gas chromatography has already been observed in a number of special regions: in elemental analysis and in the analysis of trace impurities and polymers. Further development of analytical reaction gas chromatography will undoubtedly lead to the formulation of new directions bordering on other areas of analytical chemistry (quantitative functional analysis, inorganic analysis, etc.). Prospective directions are those related with the analytical use of isotropic methods (see [33], for example) and with the selection and synthesis of complex-forming phases [33-35].

LITERATURE CITED

1. V. G. Berezkin and O. L. Gorshunov, Usp. Khim., 34:1108 (1965).
2. A. T. James and A. J. P. Martin, Biochem. J., 50:679 (1952).
3. R. L. Martin, Anal. Chem., 32:336 (1960).
4. W. H. T. Davison, S. Slaney, and A. L. Wragg, Chem. Ind. (London), 1356 (1954).
5. J. Haslam and A. R. Jeffs, J. Appl. Chem. (London), 7:24 (1957).
6. J. Voigt, Kunststoffe, 54:2 (1963).
7. J. Janak, in "Gas Chromatography 1960" (R. P. W. Scott, ed.), Butterworths, London, 1960, p. 387 [Russian translation: IL, Moscow, 1964, p. 497].
8. A. I. M. Keulemans and S. G. Perry, Nature, 193:1073 (1962).
9. F. G. Carpenter, Anal. Chem., 34:66 (1962).
10. R. L. Bond, W. J. Mullin, and F. J. Pinchin, Chem. and Ind. (London), 1902 (1963).
11. E. Bayer, Angew. Chem., 69:732 (1957).
12. J. T. Kung, J. E. Whitney, and J. C. Cavagnol, Anal. Chem., 33:1505 (1961).
13. F. Drawert, in "Gas Chromatographie 1963," ec. H. P. Angelé and H. G. Struppe, Akademie Verlag, Berlin, p. 335.
14. F. Drawert, in "Gas Chromatographie in 1961," Moscow, Gostoptekhizdat, 1963, p. 10.
15. A. I. M. Keulemans and H. H. Voge, J. Phys. Chem., 63:476 (1958).
16. R. Rowan, Anal. Chem., 33:658 (1961).
17. C. S. G. Phillips, Gas Chromatography, Butterworths, London, 1956 [Russian translation: IL, Moscow, 1958].
18. A. A. Zhukovitskii and N. M. Turkel'taub, Gas Chromatography, Moscow, Gostoptekhizdat, 1962.
19. J. E. Hoff and E. D. Feit, Anal. Chem., 36:1002 (1964).
20. E. Gil-Av and Y. Herzberg-Minzly, J. Chromatog., 13:1 (1964).
21. J. Franc and J. Jokl, Chem. Listy, 52:276 (1958).
22. M. L. Parsons, S. N. Pennington, and J. M. Walker, Anal. Chem., 35:842 (1963).

23.   J. T. Walsh and C. Merritt, Anal. Chem., 32:1378 (1960).

24.   B. Casu and L. Cavalotti, Anal. Chem., 34:1514 (1962).

25.   P. Chovin, J. Lebbe, and H. Moureu, J. Chromatog.,
      6:365 (1961).

26.   B. J. Gudzinowicz and W. R. Smith, Anal. Chem., 35:465
      (1963).

27.   U. Schwenk, H. Hachenberg, and M. Förderreuther, Brenn-
      stoff-Chem., 42:295 (1961).

28.   K. Porter and D. H. Volman, Anal. Chem., 34:748 (1962).

29.   V. G. Berezkin, A. E. Mysak, and L. S. Polak, Izv. Akad.
      Nauk SSSR, ser. khim., 1871 (1964).

30.   K. Tesarik, in "Gas Chromatographie 1965," ed. H. G.
      Struppe and D. Obst, Akademie Verlag, Berlin, 1965; Sup-
      plement, p. 89.

31.   G. E. Green, Nature, 175:422 (1955).

32.   V. G. Berezkin, V. P. Pakhomov, and V. R. Alishoev,
      Khim. i Tekhn. Topliv i Masel, No. 8, 1967.

33.   M. Senn, W. J. Richter, and A. L. Burlingame, J. Am. Chem.
      Soc., 87:680 (1965).

34.   M. E. Bednas and D. S. Russel, Can. J. Chem., 36:1272
      (1958).

35.   A. N. Genkin and B. I. Boguslavskaya, Neftekhim., 5:897
      (1965).

Chapter II

Some Problems in the Theory of Analytical Reaction Gas Chromatography

In analytical reaction gas chromatography chemical reactions and chromatographic separation are used simultaneously or in sequence. Therefore, the processes of analytical reaction chromatography are characterized not only by the usual parameters for chromatographic separation (number of theoretical plates, retention values, etc.), but also by the parameters of the chemical reaction (rate constant, activation energy, reaction orders, etc.).

Therefore, it is expedient to explain the specific features of the compounds being analyzed in chromatographic reactor columns and the influence of the reactors on the diffusion of the chromatographic zone and on the change of the retention time for the reacting and nonreacting compounds.

Depending on the actual conditions of the experiment, the different parameters related both with the chemical reaction and with the chromatographic separation can be determining factors. The proper selection of these parameters insures the success of the analysis. Thus when nonvolatile compounds are characterized by the "chromatographic spectrum" of their pyrolysis products (see [1], for example) particular attention must be given to the temperature and duration of the pyrolysis—factors which determine the degree of conversion and the "presence" of the products which are formed—as well as on the diffusion of the chromatographic zones in relation to the volume of the volatile products and the time that they enter the column.

In the kinetic variation of the analysis, in which the identification is carried out on the basis of the difference in the rate constants of the compounds being analyzed [2], the main emphasis must be placed on the kinetic rule of the chemical conversion of the compounds being eluted in the column.

In using a flow-through reactor, which is the most widespread in reaction gas chromatography, it is convenient to consider two possibilities: 1) the compound being analyzed reacts in the reactor column; 2) it passes through the reactor column without any chemical change.

CHEMICAL REACTIONS

IN CHROMATOGRAPHIC REACTORS

The specific kinetic features of chemical reactions in a chromatographic reactor column were first studied by Roginskii, Yanovskii, and Gaziev [3]. They showed that the kinetic features of chromatographic reactors are so unusual that it is necessary to consider the specific chromatographic reaction conditions, which differ considerably from the static or dynamic conditions. Under chromatographic conditions the chemical reaction takes place simultaneously (in conjunction) with the chromatographic separation of the reaction products, which results in the following features for the process [3-5]:

1) only the starting compound is present in the reaction zone as a result of the separation;

2) reversible reactions can be carried out (e.g., of the type $A \rightleftharpoons B + C$) predominantly in one direction, which makes it possible to obtain a product yield higher than the equilibrium value which could be obtained under static conditions;

3) the selectivity of the process is increased, the temperature for carrying out the reaction is decreased, side reactions are avoided, etc.;

4) the kinetic rules are simplified (e.g., the products, which may be poisons, promoters, etc., are almost absent in the reaction zone);

5) new possibilities are uncovered for studying the initial states of the function of a catalyst.

In the studies of Roginskii et al. [6-11] the application of chromatographic methods to the study of catalytic reactions was studied in detail.

The specific features of carrying out reactions under chromatographic conditions should be taken into account when using the method which was suggested by Emmett et al. [12] for the rapid evaluation of catalysts by periodically introducing a reagent into the carrier gas flow.

In studying the chemical reactions of the compounds being analyzed in a chromatographic column it is convenient to separate four cases: fast irreversible chemical reactions, slow irreversible chemical reactions, fast reversible chemical reactions, and slow reversible chemical reactions.

If a fast irreversible chemical reaction takes place in the reactor-column, then even at the very beginning of the reactor a new compound (or compounds) is formed – in which case the concentration zone of the newly formed volatile compounds is similar in shape to the zone of the original reacting compound. In this case it is necessary to take into account the possible changes in concentration due to the chemical reaction (e.g., when two moles of the product are formed per mole of the original compound) and the possible changes in the band width associated with the fact that, in the general case, the Henry coefficients of the reaction product and the reagent differ from one another.

Thus the case of a fast, irreversible reaction is basically analogous to the usual process for the separation of stable compounds in a chromatographic column. Therefore for fast reactions, the factor which usually limits the analysis is the diffusion of the zone for the products and their separation, to which particular attention should be paid.

In our opinion it is convenient to characterize fast reactions by the following arbitrary criteria: The degree of conversion of the starting compound in the reactor at a distance equal to the height equivalent to a theoretical plate (H) should already be at least 99%. Thus for fast reactions of the first order the following inequalities should be valid:

$$\frac{kHG}{u} > 2.3 \log \frac{c_0}{c}; \qquad \frac{kHG}{u} > 4.6, \tag{1}$$

where k is a rate constant for the chemical reaction; G is the Henry coefficient; u is the linear velocity of the carrier gas; c_0 and c are the initial and final concentrations, respectively.

For typical conditions (e.g., $H \simeq 0.1$ cm, $G = 10$, $u = 1$ cm/sec) the time t required to displace the zone by the distance H is equal to

$$t = \frac{HG}{u} = \frac{0.1 \text{ cm} \cdot 10}{1 \text{ cm/sec}} \text{ and } k > 5 \text{ sec}^{-1}.$$

Consequently, according to our definition the reactions with $k > 5$ sec^{-1} can be considered as fast reactions.

In the case of slow irreversible reactions, the zones which are formed in the chromatographic columns for the starting compound and the reaction products are determined by both the chromatographic and the kinetic characteristics for all the components of the reaction. For example, for the reaction $A \rightarrow B$ the shape of the zone for the starting compound and the product changes as the result of two processes: the usual chromatographic diffusion in the course of the displacement of the zone along the column, and the chemical reaction during which the concentration of the starting compound in the chromatographic zone decreases and the amount of the reaction product in the column increases.

The quantitative rules for the formation of zones for the reaction $A \rightarrow B$ were studied by Kallen and Heilbronner [13] on the basis of the concept of theoretical plates [14, 15] within the framework of nonideal linear chromatography. According to Kallen and Heilbronner, the following equation was obtained for the chromatographic zone of the initial reacting compound

$$M_N^A = \frac{m_0 \exp\left[-(N - \mu_A^* V)^2/2N\right]}{\sqrt{2\pi N}} \exp\left(-\bar{k}V\right), \qquad (2)$$

where M_N^A is the amount of substance A in the following N plates of the column, m_0 is the starting amount of substance A, N is the number of theoretical plates, μ_A^* is the coefficient of proportionality in the equation $C_{G,N}^A = \mu_A^* \cdot M_N^A$ ($C_{G,N}^A$ is the concentration of substance A in the gas phase in the last plate N), $\bar{k}$ is equal to k/w (k is the rate constant of the first-order reaction $A \rightarrow B$, and w is the volumetric flow rate of the carrier gas), and V is the volume of the carrier gas.

The retention volume (corresponding to peak maximum) of the reacting starting compound V'_{RA} can be calculated from the condition $\partial M^A_N / \partial V = 0$:

$$V'_{RA} = V_{RA} \left[1 - \frac{\bar{k}}{\mu^*_A} \right], \tag{3}$$

where V_{RA} is the retention volume of component A in the absence of a chemical reaction. Keeping in mind that $\mu^*_A = N/V_{RA}$ and $\bar{k}w = (1/t)2.3 \log (c_0/c)$, we can evaluate the term $1 - \bar{k}/\mu^*_A$ for the case when $N = 10^3$ and the reaction goes to 90% completion. In this case,

$$1 - \frac{\bar{k}}{\mu^*_A} = 1 - \frac{kt}{N} = 1 - \frac{2.3}{10^3} \approx 1.$$

Thus for effective columns and for reactions that are not too fast V'_{RA} equals V_{RA}.

On the basis of equation (2) it can be shown that the following relationship is valid for the ratio of the peak width of the reacting compounds $(\Delta V'_r)$ and the zone width of the same compound in the absence of reaction (ΔV):

$$\frac{\Delta V'}{\Delta V} = \frac{\sqrt{\left(\frac{V_{RA}\bar{k}}{N} \right)^2 + \frac{2}{N} \ln 2}}{\sqrt{\frac{2 \ln 2}{N}}}. \tag{4}$$

Under the usual experimental conditions, when, say $N = 1000$ and the degree of conversion is equal to 0.9, the term $(V_{RA}\bar{k}/N)^2$ will be small compared with $(2/N) \ln 2$. This can be seen from the following calculation:

$$\left(\frac{V_{RA}\bar{k}}{N} \right)^2 = \left(\frac{2.3 \log c_0/c}{N} \right) = \frac{2.3}{N} \approx 16 \cdot 10^{-6}; \quad \frac{2 \ln 2}{N} = 1.4 \cdot 10^{-3} \cdot$$

Consequently, $V'_R = V_R$.

The diffusion of the chromatographic zones for the reacting compound under conditions of real nonequilibrium chromatography was studied experimentally by Berezkin et al. [16]. In order to explain whether there is a diffusion of the chromatographic zones for the reacting compounds, specifically the dependence of the peak width (η) to the time of its presence in the column (t), was studied simultaneously for the peaks of the nonreacting and reacting compounds [17]. It was assumed that in the case when the points $\eta_r(t_r)$ for the reacting compound deviate from the "calibration" curve $\eta = \eta(t)$ which was obtained for the nonreacting compound it would be possible to give a quantitative evaluation of specific diffusion of the peak due to the chemical reaction; in the opposite case it would be possible to make a conclusion as to the absence of a specific effect of the reaction on the diffusion of the chromatographic peak.

The dependence of the peak width η to the retention time is given in Fig. 1 for various hydrocarbons and carrier gas flow rates (31-65 ml/min). No specific diffusion of the zone for the reacting compound (isoprene) was observed (for 5-16 liters/mole-h and a 30-70% conversion).

The equation for the chromatographic zone of the reaction product B(A → B) was obtained in [13]:

$$M_N^B = \frac{m_0 \bar{k} \exp\left[(\lambda - 1)\mu_B^* V\right]}{\lambda^{N+1} (\mu_A^* - \mu_B^*)} \int\limits_{\lambda\mu_B^* V}^{\lambda\mu_A^* V} \frac{\exp\left[-(t - N)^2/2N\right]}{\sqrt{2\pi N}} \, dt, \qquad (5)$$

where M_N^B is the amount of substance B in the last plate of the column N; μ_B^* is the proportionality coefficient in the equation $c_{G,N}^B = \mu_B^* M_N^B$ ($c_{G,N}^B$ is the concentration of the substances B in the gas phase in the last plate N), $\lambda = (1 + k)/(\mu_A^* - \mu_B^*)$; and t is the time.

On the basis of the equations some typical problems were solved using an IBM-650 computer, and the results were plotted [13].

The solution of the more general case when a homogeneous first-order reaction in the gas phase and a heterogeno-catalytic reaction of the first order with respect to the surface concentration of the adsorbed reacting material take place simultaneously in the chromatographic reactor was described by Roginskii and Rozental [18].

In this case the distribution of the concentration for the reacting substance along the column should satisfy the following system of equations:

$$\left.\begin{array}{l} \dfrac{\partial p}{\partial t} + \dfrac{\partial a}{\partial t} + u \dfrac{\partial p}{\partial x} + k_1 p + k_2 a = 0, \\[2mm] \dfrac{\partial a}{\partial t} = k_3 p - (k_2 + k_4)\, a, \end{array}\right\} \qquad (6)$$

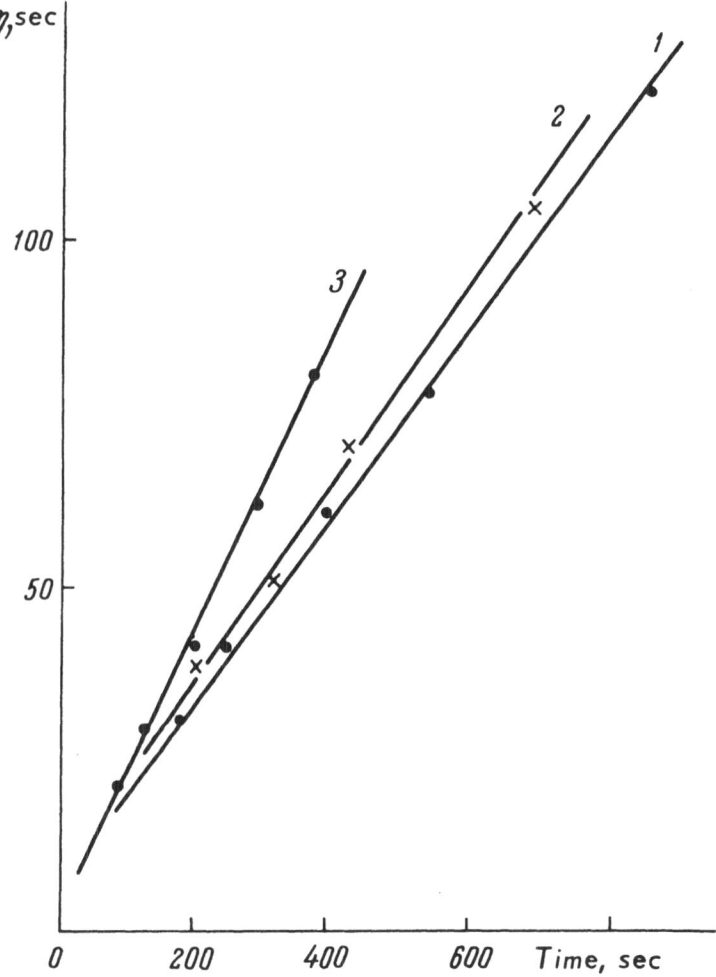

Fig. 1. Dependence of the peak width on the retention time for a mixture of hydrocarbons (in order of their elution from the column): isopentane, pentene-1, isoprene, cis-piperylene, n-heptane. Carrier gas velocity: 1) 31 ml/min; 2) 42 ml/min; 3) 65 ml/min. Column: 450 x 0.4 cm, packed with a column material containing phosphate and maleic anhydride (4:1) on a diatomaceous earth type support; column temperature, 70°C.

where p is the number of molecules of reacting gas in a unit volume of the gas phase (proportional to the partial pressure of the reacting gas); x is the distance along the column from the point of introduction of the gas into the column; a is the number of adsorbed molecules of the reacting gas corresponding to a unit volume of the gas phase; u is the carrier gas linear velocity in the column; k_1, k_2, k_3, k_4 are the rate constants for the homogeneous reaction, heterogeneous catalytic reaction, adsorption, and desorption, respectively.

The partial pressure of the reacting gas up to its entrance into the column is a known function of the time $\xi(t)$, in which case $\xi(t) = 0$ at $t = 0$ and $t - x/u \gg t_B$. At the inlet to the column $\xi(t) = p(x, t)$ where $x = 0$.

If the duration for the addition of the reacting gas into the column-reactor (t_B) is close to zero, then the following solution is obtained for the chromatographic zones of the reacting substance:

$$p(x,\ t) = \frac{2m_0 k_3 k_4 x \exp\left[(k_2 - k_1 - k_3 + k_4)\dfrac{x}{u} - (k_2 + k_4)\,t\right] J_1(z)}{uz}, \qquad (7)$$

where $z = 2\sqrt{k_3 k_4 x/u(t - x/u)}$; m_0 is the starting amount of the reacting gas; $J_1(z)$ is the first-order Bessel function of type one for the imaginary amplitude. In the range of large values for the amplitude, $J_1(z) = e^z/\sqrt{2\pi z}$. The retention time of the reacting compounds, for values of x/u which satisfy the condition $x/u > 3(k_2 + k_4)/k_3 k_4$, is expressed by (8) [18, 19]:

$$t_{max} = \frac{x}{u} + \frac{k_3 k_4 x}{2u\,(k_2 + k_4)^2}\left[1 + \sqrt{1 - \frac{3u\,(k_2 + k_4)}{k_3 k_4 x}}\right]. \qquad (8)$$

It follows from (8) that the homogeneous reaction has no effect on the retention time.

The case when $k_2 \ll k_4$ is of great interest. Expanding the factor $k_3 k_4 x/2u(k_2 + k_4)^2$ in a series with respect to k_2, we get the following expression for t_{max}:

$$t_{max} = \frac{x}{u} + \frac{x}{u}\frac{k_3}{k_4} - \frac{2x}{u}\frac{k_3}{k_4}\frac{k_2}{k_4}. \qquad (9)$$

Apparently in practice a change in t_{max} for the reacting substance can be noticed only for very fast reactions or for columns with a very small efficiency. The result that was obtained agrees with the solution found in [13].

The characteristic rules for the kinetics of reaction of different orders (n ≠ 1) for impulse chromatography were first studied by Gaziev, Filinovskii, and Yanovskii [21]. In this work the decreases in the concentration of the starting reacting compound was studied for reactions of zero, first, and second orders and for impulses of different forms, assuming that the reaction takes place under the conditions of ideal linear chromatography.

The starting system of equations includes the equation for material balance, for the rate of the chemical reaction, and for the Henry adsorption isotherms:

$$\varkappa q \frac{\partial c}{\partial t} + q(1 - \varkappa) \frac{\partial a}{\partial t} + w \frac{\partial c}{\partial x} + q(1 - \varkappa) Q = 0, \tag{10}$$
$$Q = ka^n, \quad a = Gc,$$

where c is the concentration of the substance in the gas flow; a is the amount of substance adsorbed by the unit volume of the reactor; q is the reactor cross section; $\varkappa$ is the fraction of the reactor cross section not occupied by packing; w is the volumetric gas flow rate; t is the duration of the reaction; x is the distance from the inlet of the reactor; Q is the rate of the chemical reaction of the n-th order; k is the rate constant; G is the Henry coefficient. Starting conditions for the developing variation are:

$$\left. \begin{array}{l} t \leqslant 0, \quad x \geqslant 0, \quad c(x, t) = 0; \\ t > 0, \quad x = 0, \quad c(0, t) = \psi(t). \end{array} \right\} \tag{11}$$

The solution of the system (10) for the given starting conditions has the form of an attenuating moving impulse. For n=1

$$\varphi(t) = c(l, t) = \psi(t - t_{sp}) e^{-b_1}, \tag{12}$$

for n ≠ 1

$$\varphi(t) = c(l, t) = \frac{\psi(t - t_{sp})}{\sqrt[n-1]{1 + (n-1)b_n \psi^{n-1}(t - t_{sp})}}, \tag{13}$$

$$t_{sp} = t_0 + t_{sp}^{corr} = \frac{\varkappa q + q(1 - \varkappa n)}{w}, \tag{14}$$

$$b = k t_{sp}^{corr} \, G^{n-1} = q \frac{(1-\varkappa)}{w} k G^n l, \tag{15}$$

where t_{sp} is the time the impulse remains in the reactor; t_0 is the dead time of the reactor (the time that the nonadsorbed component is present in the reactor); b is the adsorption coefficient; l is the column length.

The amount of conversion of the substance, α, can be calculated from Eq. 16:

$$1 - \alpha = \frac{\int_0^\infty \varphi(t)\, dt}{\int_0^\infty \psi(t)\, dt}. \tag{16}$$

The amount of conversion of the reacting compound can be measured with sufficient accuracy from the decrease in the area of its chromatographic peak.

In the case of a first-order reaction, the degree of conversion is not sensitive to the form of the impulse, but for reactions of zero and second orders it depends on the form of the impulse at the inlet to the reactor.

The fact that in first-order reactions the degree of conversion is independent of the form of the impulse under the conditions of gas chromatographic separation was verified experimentally by Berezkin et al. [22]. In their work, the degree of conversion in the reaction of maleic anhydride with butadiene was determined from the area of the peaks for the reacting compounds and from the height of the corresponding peaks. In Fig. 2 the ratios of the height for the chromatographic peaks before and after the reactor (h_1/h_2) are plotted on the abscissa and the ratios of peak areas (S_1/S_2) are plotted along the ordinate. Fig. 2 shows that in both cases the same results are obtained. Therefore the degree of conversion can be determined as the ratio of the peak areas, which is convenient from the experimental standpoint.

The principal features of reversible reactions under chromatographic conditions, as has already been noted, were studied by Roginskii et al. [3].

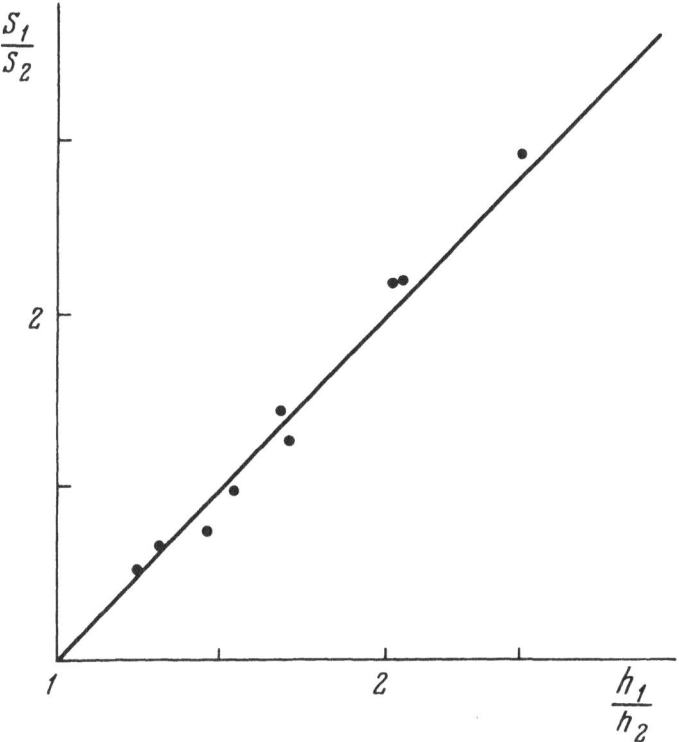

Fig. 2. The function $S_1/S_2 = f(h_1/h_2)$ for butadiene at 50.5, 60, and 74°C.

For reactions of the type $A \rightleftharpoons B + C$ for the impulsed addition of substance A (or a mixture $A + B + C$) into the zone of A, the substances B and C are constantly formed and removed from A and from one another (for different rates of movement of the compounds A, B, and C along the column reactor), which creates unique "sections" or "protuberances" behind or in front of the zone of the reacting substance A, respectively. The separation of the reacting compounds and of the reaction products makes it possible to carry out a reversible reaction in a single predominant direction, insuring that yields of the desired product will be obtained which are significantly higher than the equilibrium yield for static conditions, etc.

Roginskii and his co-workers carried out detailed investigations on the development of different variations of chromatographic conditions for studying the kinetics of complex catalytic reactions,

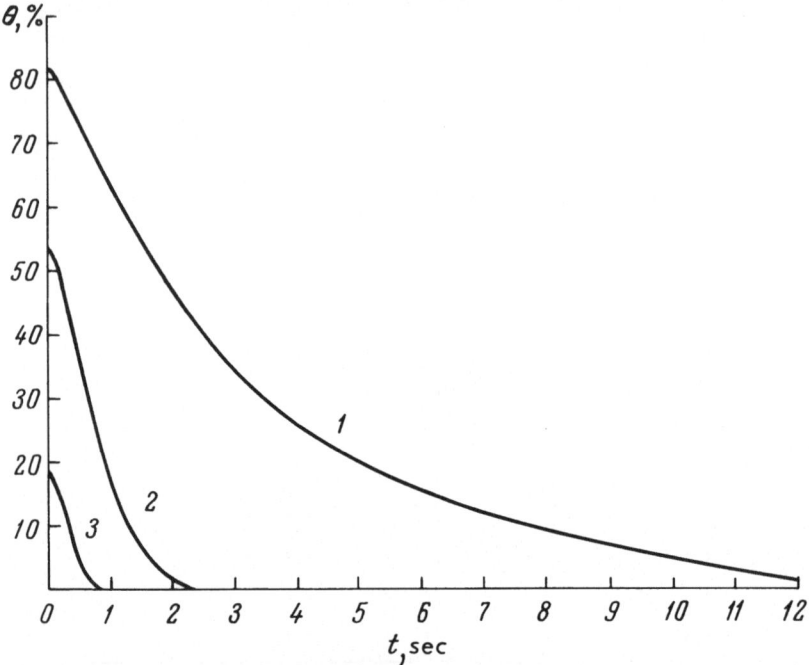

Fig. 3. The consumption of the starting compound with time ($R_C - R_A = 6$ cm/sec)
[23]: 1) K = 0.001; 2) K = 0.01; 3) K = 0.1.

and on the possibility of applying reactions under chromatographic
conditions in chemical technology. Some questions of the theory
for reversible chemical reactions were also studied by Magee [23].
If the equilibrium is established instantaneously, then, assuming
that the chromatographic process takes place under the conditions
for ideal linear chromatography, for the reaction $A \rightleftharpoons B + C$ the concentration zone for components A, B, and C should satisfy the following equations:

$$\frac{\partial c_A}{\partial t} + \frac{\partial c_B}{\partial t} = - R_A \frac{\partial c_A}{\partial x} - R_B \frac{\partial c_B}{\partial x} , \tag{17}$$

$$[c_B^2 + Kc_B] \frac{\partial c_A}{\partial t} - Kc_A \frac{\partial c_B}{\partial t} = - \frac{\partial c_A}{\partial x} [R_A c_B^2 +$$
$$+ KR_C c_B] + KR_C c_A \frac{\partial c_B}{\partial x} , \tag{18}$$

where c_A, c_B, and c_C are the concentrations of compounds A, B, and C, respectively; $K = c_B c_C/c_A$ is the equilibrium constant; R_A, R_B, and R_C are the rates of motion along the reactor for the compounds A, B and C, respectively.

The limiting conditions in this case can be written as

$$
\left.
\begin{aligned}
c_A(x, 0) + c_B(x, 0) &= D \exp(-a^2 x^2), \\
c_A(x, 0) &= \frac{1}{K} c_B^2(x, 0), \\
D &= c_A(0, 0) + c_B(0, 0).
\end{aligned}
\right\}
\qquad (19)
$$

In Magee's study this system of equations was solved by using a computer for the case when $R_A = R_B$. The results that were obtained show the possibilities of using chromatographic conditions for reversible reactions.

The dependence of the amount of unreacted starting reagent (θ) on the time of the reaction (t) is given in Fig. 3 for three different equilibrium constant values when $R_C - R_A = 6$ cm/sec. It follows from these data that even when the equilibrium constant is small, ($K = 0.001$), the degree of conversion after 10 seconds already exceeds 95%. The degree of conversion depends on the difference in the rates of movement of the products and of the starting reagent.

The plots in Fig. 4 characterize the time (t) which is required for obtaining a certain degree of conversion as a function of $1/(R_C - R_A)$ when $K = 0.1$.

Fig. 5 plots the dependence of the time (t) required for obtaining a certain degree of conversion as a function of the equilibrium constant (K).

Magee [23] also showed that the reaction, when it is carried out under chromatographic conditions, ($R_A - R_C = 6$ cm/sec) can almost be completed even if the equilibrium constant $K = 2 \cdot 10^{-7}$. However if $K = 0.1$ the reaction will take place even if $R_A - R_C$ is only $3 \cdot 10^{-4}$ cm/sec.

It should however be noted that in practice, due to the diffusion of the chromatographic zones, the characteristics of the process may change significantly if the chromatographic column being used does not have adequate efficiency. For example let us evaluate the number of theoretical plates (N) required to separate two compounds ($R_A - R_C = 3 \cdot 10^{-4}$ cm/sec and $R_A/R_C = 1.006$), with the

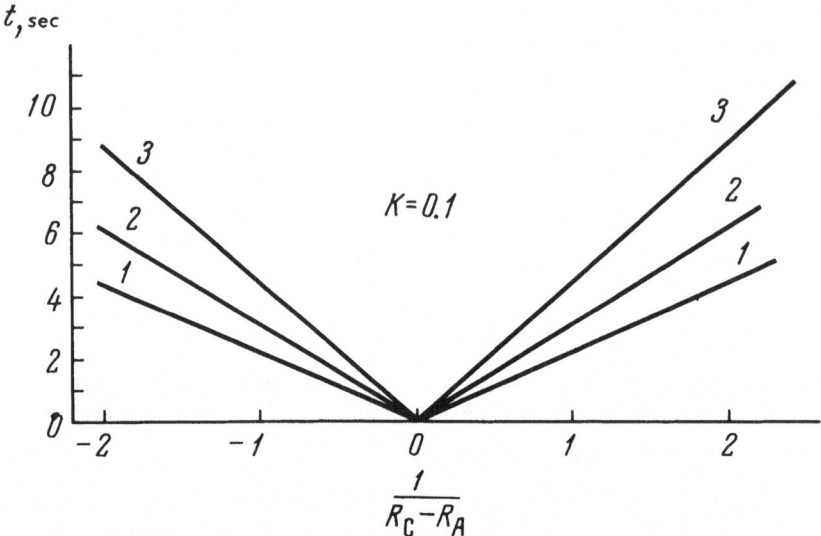

Fig. 4. Time required to obtain a certain degree of conversion, $R_C - R_A$ as a function of $1/(R_C - R_A)$ [23]: 1) 90%; 2) 95%; 3) 99% conversion.

assumption that $K_B = 0.1$. According to Reich's equation [20]

$$N = \ln \frac{4}{1 - K_B} \left(\frac{r + 1}{r - 1}\right),$$

where r can be assumed equal to R_A/R_C; therefore,

$$N = \frac{\ln 4.4}{9} \cdot 10^6 = 1.6 \cdot 10^5.$$

To obtain such a high efficiency it would be necessary to use capillary columns.

The theory of slow reversible chemical reactions of the type
$A \rightleftharpoons B$ which take place in a chromatographic column was studied
by Klinkenberg [24] who studied the effect of the reaction rate on
the retention time and on the shape of the elution peak. An analysis
of a similar problem was also given by Keller and Giddings [25].
We have found that broadening of the peaks for some components,
which can be observed in chromatographic analysis in highly efficient columns, may be explained by the occurrence of slow reversible
reactions during the process of chromatographic separation [24, 26].

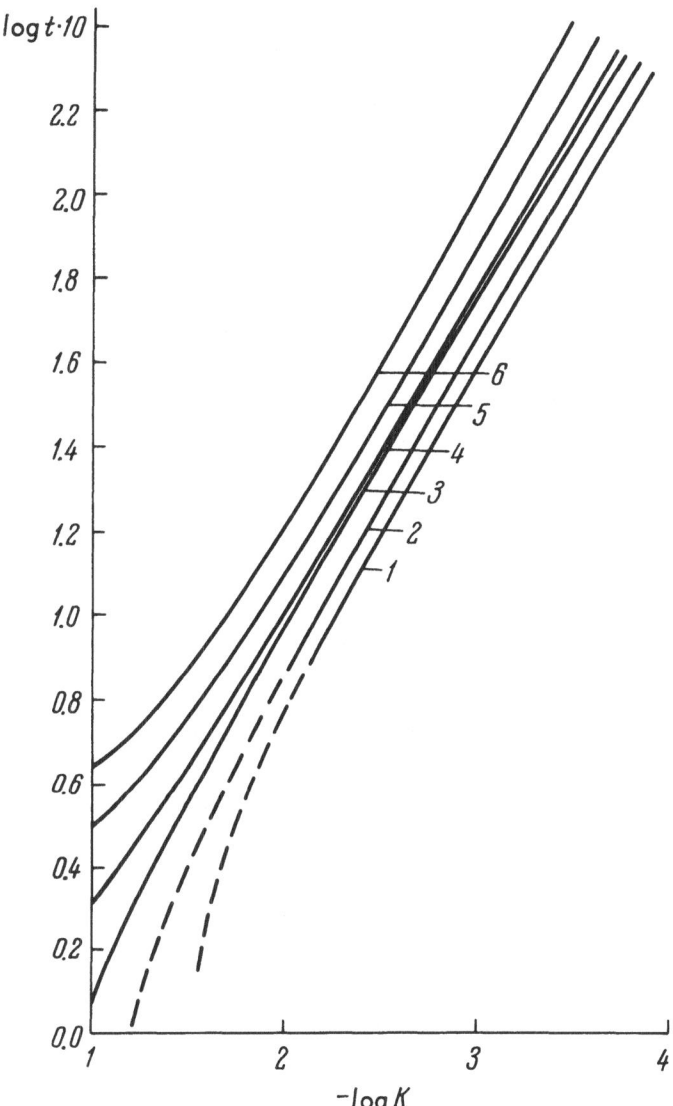

Fig. 5. Time required to obtain a certain degree of conversion as a function of the equilibrium constant $(R_A - R_C = 6 \, \text{cm/sec})$ [23]: 1) 70%; 2) 75%; 3) 83%; 4) 85%; 5) 90%; 6) 95% conversion.

The kinetic rules that were studied above for chemical reactions which take place in the chromatographic columns must be taken into account in the analytical practices. They can be used successfully together with purely chromatographic characteristics for the identification of the reacting compounds and for the determination of their starting concentrations in the samples being analyzed (chromatographic kinetic method of analysis).

ANALYTICAL APPLICATION OF THE KINETIC METHOD IN GAS CHROMATOGRAPHY

The analytical methods which are based on the measurement of the kinetic characteristics of the reaction (rate constant, activation energy, etc.) are called kinetic methods. The kinetic methods of analysis can be used both for the identification of the sample components or for determining their concentration in the original sample. At the present time the kinetic methods of analysis are widely used in the analytical chemistry of inorganic compounds [27]. Chromatographic and, specifically, gas chromatographic methods greatly expand the possibilities of the kinetic method as compared with the chemical reaction being carried out without separation of the mixture being analyzed into the individual components. The use of the kinetic method in conjunction with chromatographic separation of the sample components makes it possible to avoid the effect of harmful side reactions, to expand the group of reactions which can be used for these determinations (in this case it is possible to lower the requirements for the selectivity of these reactions), and to obtain, in a single experiment, data on the kinetics of all of the sample components which are of interest to the investigators.

Gil-Av and Herzberg-Minzly [2] suggested using data both on the retention volume (chromatographic characteristics) and on the rates of chemical reaction of the substances being analyzed with the stationary phases (kinetic characteristics) for their identification. This suggested method was called the partial subtraction method by analogy with the earlier known subtraction method based on the complete absorption of certain sample components by a selective reagent. However, it would be more proper to call the suggested method the gas chromatographic-kinetic method since, for the identification of the unknown compounds, their kinetic characteristics are used (absolute or relative values).

Gil-Av and Herzberg-Minzly studied as an example the first-order reaction between conjugated dienes (the subtracted components) and chloromaleic anhydride stationary liquid phase. The reaction products are nonvolatile and remain in the column. The rate of the chemical reaction can be calculated by the equation

$$\frac{dm}{dt} = -km\sigma,$$

(20)

where m is the total amount of the reacting component, t is the reaction time, σ is a factor by which m must be multiplied in order to obtain the amount of compound dissolved in the stationary liquid phase, and $\sigma = V_R'/V_R$ (V_R and V_R' are the uncorrected and true retained volume, respectively). For compounds with a relatively large retention volume, $\sigma \approx 1$. Taking into account that the reaction time equals V_R/w (w is the carrier gas flow rate), by the integration of Eq. (20) the following kinetic equation can be obtained:

$$\log \frac{m_0}{m} = 0.4343 \, k \frac{V_R}{w} \sigma.$$

(21)

The authors indicate that for the usual method of subtraction (i.e., for practically complete absorption), $\log m_0/m \geq 5$. In the kinetic method this value can change within the limits of 0.04–1.0.

Thus the rates of the chemical reaction used in the kinetic method change over a broad range. Therefore, a significant change in the values of the rate constant can be obtained by controlling the experimental conditions: for example, by changing the temperature, introducing a catalyst, using a solution of the reagent instead of the pure compound, etc.

Gil-Av and Herzberg-Minzly [2] studied the chromatographic separation and reaction with chloromaleic anhydride at 40°C for the following dienes: 1,3-pentadiene, 2,4-hexadiene, 1,3,5-hexatriene, and cyclopentadiene. Firebrick C-22 coated with chloromaleic anhydride stationary phase (2:1, w:w) was used as the column packing. In using a single column with this stationary phase, the reaction time was assumed to be equal to the retention time of the reacting compound. In some experiments multiple columns were used: the first column served only as an analytical column containing an inert liquid phase, while the second was the reaction column. In this case, the duration of the reaction was determined from the

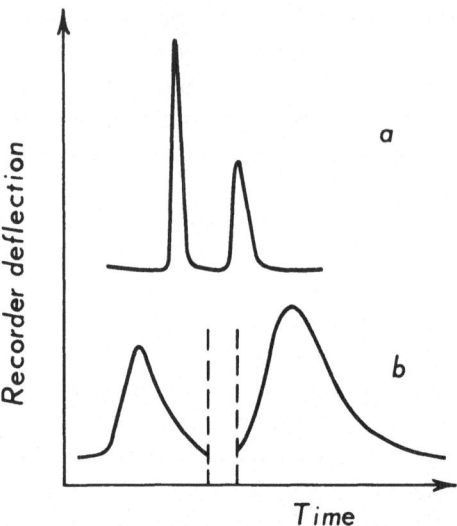

Fig. 6. The chromatogram for technical 1,3-
pentadiene on a double-column system (first
peak, trans-isomer; second peak, cis-isomer)
[2] Carrier gas (He) flow rate: a) 77; b) 11.4
ml/min.

equation

$$V_R = jwt,$$

where V_R is the corrected retention volume for the chloromaleic
anhydride reaction column; j is the compressibility correction factor.

Gil-Av and Herzberg-Minzly [2] used the kinetic method to
determine the structure of conjugated dienes. The rate of chemical
reaction between the dienes and the dienophyl reagents, e.g., chloromaleic anhydride) depends greatly on the structure of the reacting
diene. The trans-isomer usually enters readily into reaction with
the dienophyl reagent, but the cis-isomer reacts much more slowly. The chromatograms for technical 1,3-pentadiene obtained on a
double-column system (in the first column a solution of silver nitrate in ethylene glycol was used as the stationary liquid phase,
while in the second chloromaleic anhydride was used) are shown in
Fig. 6 for two different flow rates (and, consequently, reaction
times). Under the given analytical conditions, technical 1,3-penta-

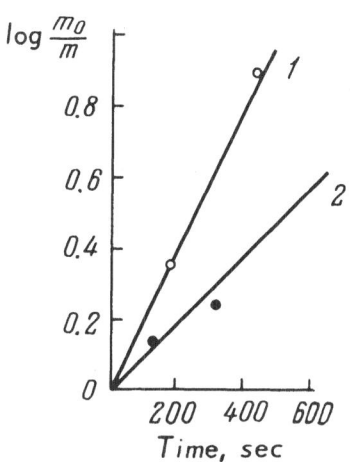

Fig. 7. Relationship between log m_0/m and the reaction time (contact) for: 1) isoprene; 2) trans-1,3-penta-diene [2]. Length of the chloromaleic anhydride column: 1-m Column tem-perature 40°C.

diene is separated into the cis- and trans-isomers. If the time of analysis is increased, the relative area of the first peak correspond-ing to the more reactive trans-isomer decreases from 60 to 31%. The trans, trans-2,4-hexadiene in a mixture with trans-1,3-hexa-diene was identified by an analo-gous method and a number of other analytical problems were also solved.

The kinetic method was also used to identify the peaks of the compounds being analyzed on the basis of the accurate measure-ment of the rate of the chemical reaction between the dienes and chloromaleic anhydride. The kinetic data for the reaction of isoprene and trans-1,3-pentadiene are shown in Fig. 7. The straight lines were drawn on the basis of reaction studies with the pure known compounds. The experimental points correspond to the data for chromatographic experiments obtained in analyzing an unknown mixture. Thus, carrying out the chromatographic analysis with two different carrier gas flow rates, it is possible to identify the reacting compounds on the basis of their reaction rates with a nonvolatile reagent in the reactor column.

It should be pointed out that, independently of the above dis-cussed publication, an analogous method was developed by Berezkin et al. [22] when studying the kinetics of bimolecular chemical reac-tions by impulse chromatography; in this case the chromatographic system incorporated two detectors, so that chromatographic and kinetic data could be obtained in a single experiment.

In the chromatographic system employed [22] the column preceded the reactor, which made it possible to study the kinetics of the chemical reaction of the compounds in a pure form without separating them first. The concentration of the reacting compo-nents was recorded by two independent detectors before and after the reaction. The use of this system made it possible to obtain, in a single experiment, values for the concentration of the reacting compounds before and after the reaction, with a known reaction time; these are the values necessary to calculate kinetic data. By changing the concentration of the second, nonvolatile component the order of the reaction with respect to this component can be determined, and the rate constant of the bimolecular reaction can be calculated. This was not taken into account by Gil- Av and Herzberg- Minzly [2], who studied the impulse reaction of a diene with chloro- maleic anhydride stationary phase only as a pseudo first-order re- action.

In [22] the kinetics for the reaction of diene synthesis with maleic anhydride was studied by impulse chromatography for iso- prene and butadiene.

Forty percent maleic anhydride was deposited in the form of a saturated solution at 45°C in tricresylphosphate on the solid sup- port. Tricresylphosphate was used as a solvent in order that: 1) the usable temperature range could be expanded (since maleic anhydride melts at 56°C and begins to sublime noticeably at 70°C); and 2) the contact time of the reacting substances could be in- creased.

The concentration of the maleic anhydride was 6.5 mole/liter; the concentration of the butadiene, introduced into the reactor in the form of a slug was about $1.5 \cdot 10^{-3}$ mole/liter. Only 0.02% of the maleic anhydride was consumed in a single experiment. In using highly sensitive detectors this value can be as low as $10^{-5}\%$, i.e., the concentration of the reagent in the stationary phase re- mains almost constant.

When the concentration of the maleic anhydride remains con- stant the reaction rate constant can be calculated according to

$$k = 2.3 \left(\log \frac{m_0}{m} \right) \Big/ c_{MA}^{r} t, \qquad (22)$$

where c_{MA} is the concentration of the maleic anhydride, and n is the reaction order with respect to the maleic anhydride. The reac- tion time was determined by the difference in the retention time of the diene before and after the reaction, taking into account the dead time of the reactor. It should be pointed out that to identify the compounds being analyzed it is better to make use of the relative values of the rate constant rather than the absolute values; it is also convenient to use as a characteristic the value of the activation energy, which is often determined with great accuracy, rather than the rate constant.

The kinetic method of analysis also makes it possible to determine the starting concentration of the reacting compounds if their kinetic equations are known. This is particularly important for the chromatographic analysis of unstable reactive substances.

The fact that it is possible to make use of kinetic rules in analytical reaction gas chromatography to calculate the starting concentrations of the reacting compounds was shown by Berezkin et al. [16] by the example of the reaction of butadiene with maleic anhydride. The authors utilized the internal standard method (see, for example, [28]). The calculation for a first-order reaction was carried out by means of

$$2.3 \log\left(\frac{S_r}{S_{st}}\right)_t = 2.3 \log\left(\frac{S_r}{S_{st}}\right)_0 - kt, \tag{23}$$

where $(S_r/S_{st})_0$ and $(S_r/S_{st})_t$ are the ratios of the peak areas of the reacting compounds and the stable standard at time 0 and t. The relative starting concentration $(S_r/S_{st})_0$ was determined either by calculation or graphically on the basis of (23). The starting concentration of divinyl, found graphically as $(S_r/S_{st})_0 = 2.72$ is very close to the actual value of $(S_r/S_{st}) = 2.66$. If it is assumed that the contact (reaction) time is inversely proportional to the carrier gas flow rate, a simplified equation can be used for the calculation:

$$\frac{\log \frac{(S_r/S_{st})_0}{(S_r/S_{st})_1}}{\log \frac{(S_r/S_{st})_0}{(S_r/S_{st})_2}} = \frac{w_2}{w_1}. \tag{24}$$

The starting concentration of divinyl calculated by this method differs from the true value only by 3-4 relative percent.

Thus the kinetic chromatographic method expands the number of reactions of different types which can be used for group identification of the compounds being analyzed, makes it possible to carry out the individual identification of the reacting components

on the basis of the measurement of the individual kinetic characteristics, and permits determination of the initial concentration of the
reacting substances by means of the known kinetic equations.

EFFECT OF THE REACTOR
ON THE CHROMATOGRAPHIC CHARACTFRISTICS
OF THE COMPOUNDS BEING ANALYZED

The effect of the reactor on the retention time and the width
of the chromatographic zone for compounds which react slowly in
the ehromatographic column (chromatographic reactor) were discussed above. However in analytical reaction gas chromatography,
reactions in nonchromatographic reactors, which are placed in
front of the column (the analysis of nonvolatile compounds, elemental analysis, etc.), are also widely used, as are fast reactions in
reactors of the chromatographic type, which are placed either after
or in front of the chromatographic column (conversion of water into
acetylene, conversion of organic compounds to carbon dioxide on
copper oxide, etc.).

In the chromatographic system the reactor is usually in
series with the chromatographic column. Therefore in studying
the effect of the reactor on the chromatographic characteristics of
the compounds being analyzed it is necessary to consider the following two cases:

1. for reactors of the nonchromatographic type – the relation
of the sample characteristics (volume, time of introduction into
the column, etc.) with the efficiency of the following chromatographic separation;

2. for chromatographic-type reactors, placed in series with
the column – the influence of their characteristics on the retention
time and the efficiency of chromatographic separation of the nonreacting components and stable reaction products.

These problems were not investigated directly; however, they
can be studied on the basis of information available in the literature
on the dependence of the efficiency of the chromatographic separation on sample characteristics, and on the basis of the theory of
separation in composite columns.

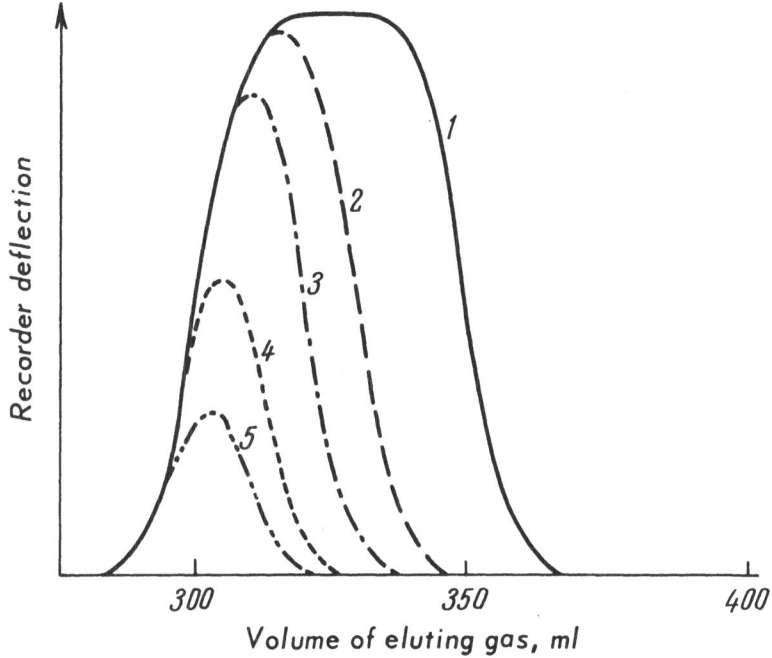

Fig. 8. The effect of the sample volume on the shape and width of the chro matographic peak [29] for the peak corresponding to $V_R = 300$ ml, $N = 2000$. Sample volumes: 1) 50 ml; 2) 30 ml; 3) 20 ml; 4) 10 ml; 5) 5 ml.

The efficiency of the chromatographic separation decreases as the sample volume increases and as the time for its introduction into the column increases.

In Fig. 8 the influence of sample volume on the width and shape of the chromatographic peak is shown [29]. It can be seen that the peak width increases noticeably as the sample volume increases, resulting in the decrease in the efficiency of the chromatographic column (decrease in the number of theoretical plates N).

It was shown by Glueckauf and van Deemter et al. [14, 30] that the limiting sample volume v_l, i.e., the maximum sample volume which can be introduced into the column without a decrease in the efficiency, can be determined from

$$v_l = \varepsilon \frac{V_R}{\sqrt{N}},$$

(25)

where ε is a constant. Keulemans [31] recommended using $\varepsilon = 0.5$ for the analysis of vapor samples, and $\varepsilon = 0.02$ for pure gas samples. Thus, in analyzing vapor mixtures, if $N = 1000$ and $V_R = 100$ ml, then $v_l = 1.5$ ml; if $N = 1000$ and $V_R = 1000$ ml, then $v_l = 15$ ml; and if $N = 10,000$ and $V_R = 100$ ml, then $v_l = 0.5$ ml.

In practice it is often necessary to analyze samples with volumes exceeding v_l, which leads to a decrease in the separation efficiency [32].

To evaluate the width of the chromatographic zone formed during separation when the column is overloaded (the volume of the sample is greater than v_l) it is convenient to represent the broad initial zones by a rectangular shape (width η_p) which is the sum of individual narrow single rectangular zones, each of which has a volume equal to v_l. Each of these single samples (with a v_l volume), after passing with the carrier gas through the column, forms a single chromatographic zone with a bell-shaped form having a width of η_B. In the first approximation the process of forming a chromatographic zone in overloaded columns can be represented as the result of the superposition (summation) of individual bell-shaped single zones which are formed during the independent chromatographic analysis of the small sample volumes which make up the η_p initial zone width. Therefore, in the case of a large sample, the width of the chromatographic zone after the column (η_0) will be equal to the sum of the widths of the initial zones (η_p) and a single chromatographic zone (η_B):

$$\eta_0 = \eta_p + \eta_B. \tag{26}$$

Keeping in mind that

$$H = \frac{L}{16}\left(\frac{\eta}{t}\right)^2 \tag{27}$$

(where L is the column length and t is the retention time), the following approximate equation can be obtained which relates the change in the height equivalent to a theoretical plate (H) to the change in the initial width of the sample:

$$\frac{H_0}{H_B} = \left(1 + \frac{\eta_p}{\eta_B}\right)^2. \tag{28}$$

Equation 28 agrees with the experimental data given by Purnell and Sawyer [33].

As an example, in Fig. 9, the influence of the sample size on the value of H is shown for c_5-c_7 n-paraffins [34]. Column efficiency decreases as the sample volume increases, and the change is greater for compounds with smaller partition coefficient.

The effects of the time of sample introduction on the efficiency of the separation were studied by Guiochon [35]. He showed that, assuming the additivity of the diffusion processes, the following equation is valid:

$$H = H_0 + H_t,$$

(29)

where H_0 is the height equivalent to a theoretical plate which would be obtained when introducing the sample at "zero" time into the column; and H_t is the contribution to the value of H due to the finite sample introduction time:

$$H_t = \gamma t^2 v^2 u^2,$$

(30)

where v is the ratio of the rate of displacement of the zone along the column to the carrier gas flow rate, t is the time of sample introduction, u is the linear carrier gas velocity, and γ is a constant which changes for different compounds within the limits of 0.005-0.07 [35]. The value of H_t, particularly during rapid analysis, is rather large. Thus, for example, for $v = 0.5$, u = 20 cm/sec, L = 100 cm and t = 0.5 sec, $H_t = 2.5$ mm.

In using slow chemical reactions, when it is impossible to decrease the sample introduction time, it is expedient to use either an intermediate concentration or to connect the reactor with the chromatographic system only after the reaction has been completed.

In a number of cases the reactor and the chromatographic column are thermostated at different temperatures and are connected with capillary tubes. It should be pointed out that the temperature of the reactor and of the connecting tubes must not be lower than the boiling point of the sample components, because otherwise condensation might occur, decreasing the efficiency of the separation. As an illustration, Fig. 10 plots the number of

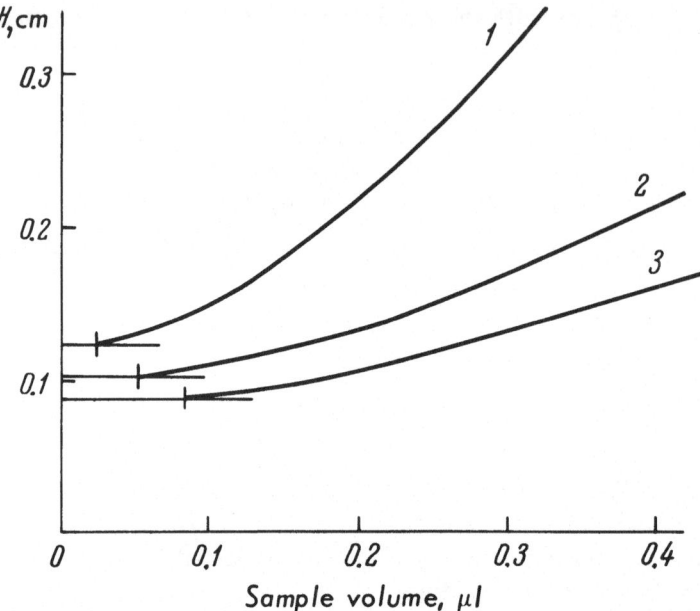

Fig. 9. The dependence of the height equivalent to a theoretical plate on the size of the sample being analyzed for a low capacity column [34]. 1) n-pentane (k' = 1.6); 2) n-hexane (k' = 4.2); 3) n-heptane (k' = 10.5). k' is the partition ratio described by the equation $k' = K(V_L/V_G)$, where K is the partition coefficient and V_L and V_G are the volume of the liquid and gas phases in the column, respectively.

theoretical plates obtained as a function of the temperature of the injection block [36]. It is evident from this figure that the efficiency of the column decreases sharply if the temperature of the sample-introduction device is lower than the boiling point of the sample components.

The rules considered above for the sample diffusion must be taken into account when selecting operating conditions for non-chromatographic-type reactors.

In chromatographic-type reactors the values of retention and diffusion of the zones for both the product and the unreacted compounds can be determined on the basis of the corresponding rules for chromatographic columns which have been studied in detail [20, 31, 34].

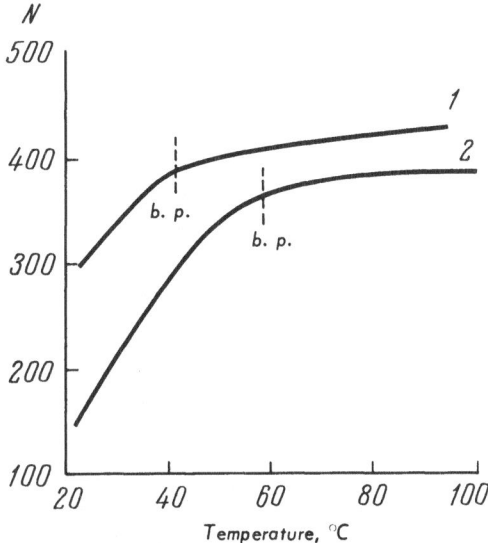

Fig. 10. The influence of the temperature of the injection block on the number of theoretical plates. Column temperature, 41°C. 1) dichloromethane (b.p. = 40°C, sample volume 3 μl); 2) acetone (b.p. = 56°C, sample volume 2 μl).

Liquid reagents (catalysts) are deposited as thin films on a solid support. Usually, the reactor temperature is either equal to or higher than the temperature of the chromatographic column, and sometimes, it is significantly higher (e.g., in the conversion of organic sample components to carbon dioxide and water). The chromatographic reactors are often much shorter than the chromatographic columns, while the diameter of the reactor and chromatographic column are usually about the same. Thus for nonreacting compounds and for the volatile products formed (fast reactions), the reactors of this type can be considered as chromatographic columns.

Chromatographic-type reactors are connected in series with the chromatographic columns, and therefore, for the determination of the chromatographic characteristics (retention time, zone width), calculation methods which have been developed for composite columns can be used [37–39].

In practice relative retention time can be calculated with the help of the formula [39]:

$$\tau_i = a\tau_{i1} + b\tau_{i2};$$

$$a = \frac{1}{1+\lambda_{1,2}}; \quad b = \frac{1}{1+1/\lambda_{1,2}}; \quad a+b=1;$$

$$\lambda_{1,2} = \frac{t_{st2}}{t_{st1}} = f_{1,2}\frac{P_0\,x\,V^0_{st2}}{P_{1,2}\,y\,V^0_{st1}},$$

(31)

where τ_{i1} and τ_{i2} are the relative retention times in columns 1 and 2; t_{st1} and t_{st2} are the retention times of the standard substance sample in columns 1 and 2; $f_{1,2}$ is a term which takes into account the pressure gradient (in connecting the columns: first, second); $P_{1,2}$ is the pressure between the columns; P_0 is the pressure at the outlet of the composite column; V^0_{st1} and V^0_{st2} are the specific retention values of the standards in columns 1 and 2 calculated per unit of weight, volume, or length; x and y are the lengths of columns 1 and 2.

The effect of the reactor on the total relative retention time is estimated for the widespread case when the length of the reactor is only a small percent of the chromatographic column length and the specific retention values for the reactor are smaller than those for the chromatographic column. Examples can be found in the selective absorption of unsaturated compounds by concentrated sulphuric acid, in the conversion of water into acetylene on calcium carbide, in the conversion of organic compounds on copper oxide, etc. Thus, in accordance with Eq. (31),

$$f_{1,2} \approx 1; \quad P_0/P_{1,2} \approx 1;$$

$$\frac{V^0_{st2}}{V^0_{st1}} < 1, \quad \frac{x}{y} < 2\cdot10^{-2} \quad \text{and} \quad \lambda_{1,2} < 2\cdot10^{-2};$$

$$a = 0.98; \quad b < 2\cdot10^{-2}; \quad \tau_i = 0.98\cdot\tau_{i1} + 2\cdot10^{-2}\cdot\tau_{i2} \approx \tau_{i1}.$$

From this evaluation it follows for the given case that the relative retention time in a reactor–column system is almost identical to the corresponding value of the chromatographic column alone ($\tau_{i1} > 0.2$).

The calculation of the absolute values of the retention volume and the width of the chromatographic peak can be carried out by means of the method developed by Vigdergauz and Gol'bert [38], which gives the width of the peak in a composite column η from the peak widths η_1, η_2 obtained separately in columns 1 and 2:

$$\eta^2 = \eta_1^2 + \eta_2^2. \tag{32}$$

Since $H = \varepsilon L(\eta/V)^2$, the height equivalent to a theoretical plate H for a composite column can also be obtained:

$$H = H\rho_1 + H_2\rho_2;$$

$$\rho_1 = \left(\frac{V_1}{V}\right)^2 \frac{L}{x}; \quad \rho_2 = \left(\frac{V_2}{V}\right)^2 \frac{L}{y}, \tag{33}$$

where H_1 and H_2 are the values of the height equivalent to a theoretical plate for columns 1 and 2, V_1 and V_2 are the individual retention volumes, V is the total retention volume, and x, y are the lengths of columns 1 and 2; $x + y = L$.

For the case considered above, when having a short reactor which retains the sample only for a very small time,

$$\rho_1 = 1; \quad \rho_2 < 1 \quad \text{and} \quad H = H_2\rho_2 + H_1 \approx H_1.$$

The diffusion of the zones for the nonretained components (the extended case for chromatographic reactors) were studied in detail by a number of investigators [40-42].

The value of H for the nonretained components is usually small. Thus, for example, according to the data published by Knox and McLaren [42] for the diffusion of ethylene in nitrogen, H_{min} is 0.7 mm for a packed column (length 5.93 m, diameter 3.8 mm) and for a capillary column (length 16 m, diameter 0.88 mm) $H_{min} = 0.34$ mm. Therefore, it is often possible to assume that $H \approx H_1$. It should be noted that the values of the retention time and efficiency H for a chromatographic column with or without a reactor, calculated according to the data of Martin [43], are almost identical.

In order to decrease the diffusion, it is convenient to utilize, whenever possible, either capillary reactors or packed columns with small diameters [44-45].

LITERATURE CITED

1. V. G. Berezkin, O. L. Gorshunov, and M. A. Geiderikh, Plast. Massy, No. 11:53 (1965).

2.   E. Gil-Av and Y. Herzberg-Minzly, J. Chromat., 13:1 (1964).
3.   S. Z. Roginskii, M. I. Yanovskii, and G. A. Gaziev, Dokl.
      Akad. Nauk SSSR, 140:1125 (1961).
4.   S. Z. Roginskii, M. I. Yanovskii, and G. A. Gaziev, (Russian
      patent) Avt. Svid. SSSR 149398 (1961); Byull. Izobr., No. 16
      (1962).
5.   G. A. Gaziev, Doctoral Dissertation, Moscow Inst. Khim. Fiz.,
      Akad. Nauk SSSR, 1965.
6.   G. A. Gaziev, O. V. Krylov, S. Z. Roginskii, G. V. Samsonov,
      E. A. Fokina, and M. I. Yanovskii, Dokl. Akad. Nauk SSSR,
      140:863 (1961).
7.   S. Z. Roginskii, M. I. Yanovskii, and G. A. Gaziev, Kinetika
      i Kataliz, 3:529 (1962).
8.   S. Z. Roginskii, M. I. Yanovskii, and G. A. Gaziev, in "Gas
      Chromatography," Proceedings of the Second All-Union
      Conference, Moscow, "Nauka" Press, 1964, p. 27.
9.   S. Z. Roginskii, É. I. Semenenko, and M. I. Yanovskii, Dokl.
      Akad. Nauk SSSR, 153:383 (1963).
10.   É. I. Semenenko, S. Z. Roginskii, and M. I. Yanovskii,
      Kinetika i Kataliz, 5:490 (1964).
11.   É. I. Semenenko, S. Z. Roginskii, and M. I. Yanovskii,
      Kinetika i Kataliz, 6:320 (1965).
12.   R. J. Kokes, H. Tobin, and P. H. Emmett, J. Am. Chem. Soc.,
      77:5860 (1955).
13.   J. Kallen and E. Heilbronner, Helv. Chim. Acta, 43:489 (1960).
14.   E. Glueckauf, Trans. Faraday Soc., 51:34 (1955).
15.   A. Klinkenberg and F. Sjenitzer, Chem. Eng. Sci., 5:258
      (1956).
16.   V. G. Berezkin, V. S. Kruglikova, and V. E. Shiryaeva,
      Neftekhimiya, 6:630 (1966).
17.   B. D. Blaustein and G. M. Feldman, Anal. Chem., 36:65
      (1964).
18.   S. Z. Roginskii and A. L. Rozental, Dokl. Akad. Nauk SSSR,
      146:152 (1962).
19.   S. Z. Roginskii and A. L. Rozental, Kinetika i Kataliz, 5:104
      (1964).
20.   A. A. Zhukhovitskii and N. M. Turkel'taub, Gas Chromatog-
      raphy, Gostoptekhizdat, Moscow, 1962.
21.   G. A. Gaziev, V. Yu. Filinovskii, and M. I. Yanovskii,
      Kinetika i Kataliz, 4:688 (1963).

22.   V. G. Berezkin, V. S. Kruglikova, and N. A. Belikova, Dokl. Akad. Nauk SSSR, 158:182 (1964).

23.   E. M. Magee, Ind. Eng. Chem. Fundament., 2:32 (1963).

24.   A. Klinkenberg, Chem. Eng. Sci., 15:255 (1961).

25.   R. A. Keller and J. C. Giddings, J. Chromatog., 3:205 (1960).

26.   T. R. Phillips and D. R. Owens, in "Gas Chromatography 1960" (R. P. W. Scott, ed.), Butterworths, London, 1960, p. 308.

27.   K. B. Yatsimirskii, Kinetic Methods of Analysis, Goskhimizdat, Moscow, 1962.

28.   V. G. Berezkin, I. A. Musaev, V. S. Tatarinskii, and P. I. Sanin, in "Gas Chromatography," INNTFKhIM, Moscow, 1966, p. 25.

29.   P. E. Porter, C. H. Deal, and F. H. Stross, J. Am. Chem. Soc., 78:2999 (1956).

30.   J. J. van Deemter, F. J. Zuiderweg, and A. Klinkenberg, Chem. Eng. Sci., 5:271 (1956).

31.   A. I. M. Keulemans, Gas Chromatography, 2nd ed., Reinhold, New York, 1959, p. 199.

32.   M. S. Vigdergauz and M. I. Afanas'ev, Neftekhimiya, 3:911 (1963).

33.   H. Purnell and D. T. Sawyer, Anal. Chem., 36:668 (1964).

34.   S. D. Nogare, and R. S. Juvet, Gas Chromatography, Interscience, J. Wiley and Sons, New York, 1962 [Russian translation: "Nedra" Press, Leningrad 1966].

35.   G. Guiochon, Anal. Chem., 35:399 (1963).

36.   F. H. Pollard and C. J. Hardy, Chem. and Ind. (London), 1145 (1955).

37.   A. A. Zhukhovitskii, M. S. Selenkina, and N. M. Turkel'taub, Zh. Fiz. Khim., 36:993 (1962).

38.   M. S. Vigdergauz and K. A. Gol'bert, Neftekhimiya, 2 : 852 (1962).

39.   V. G. Berezkin and I. V. Sidorova, Neftekhimiya, 3 : 144 (1963).

40.   K. V. Alekseeva, A. A. Zhukhovitskii, and N. M. Turkel'taub, Khim. i Tekhnol. Topliv i Masel, No. 4:60 (1962).

41.   J. C. Sternberg and R. E. Poulson, Anal. Chem., 36:1492 (1964).

42.   J. H. Knox and L. McLaren, Anal. Chem., 35:449 (1963).

43.   R. L. Martin, Anal. Chem., 32:336 (1960).

44.   I. Halasz and E. Heine, Nature, 194:971 (1962).

45.   V. G. Berezkin, A. T. Svyatoshenko, and L. V. Klement'evskaya, Zh. Anal. Khim., 21:1367 (1966).

Chapter III

Chromatographic Systems in Analytical Reaction Gas Chromatography

In analytical reaction gas chromatography, along with the usual elements of chromatographic systems (columns, detectors, switching devices, etc.), new elements are used – chemical reactors in which all or some of the sample components undergo certain chemical conversions. The use of such elements results in new chromatographic systems which should be studied in greater detail since they represent important parts of gas chromatography. The great variety of important practical analytical problems results in an equal variety of possible reaction chromatographic systems.

The most typical simple systems of analytical reaction gas chromatography are illustrated in Table 1. For comparison, the usual schematic of a gas chromatography without a reactor is also given in this table (system 0).

In practice the chromatographic column serves only in a few – presently very rare – cases simultaneously as the reactor (system 4); usually the function of the chromatographic column and of the chemical reactor are separated.

Systems 1, 3, and 5 are the most widespread in analytical reaction gas chromatography; system 7 is used rather rarely. The application of systems 2, 4, and 6, at the present time, is mentioned only in individual studies.

TABLE 1. Chromatographic Systems in Analytical Reaction Gas Chromatography

No.	Simplified schematic of the chromatographic system	Reference (example)
0		[1]
1		[2]
2		[3]
3		[4]
4		[5]
5		[6]
6	a b	[7, 8]
7		[9]
8	a b	[10]

Explanation of the symbols: ◯, sample introduction device; ⊠, reactor; 〰, chromatographic column; ⬚, detector with sensing and reference sides; △, registering and recording device.

Generally, each individual system can be used successfully for the solution of various analytical problems, although a particular system is preferred for any given application. Thus for example, system 1 is predominantly used in the analysis of polymers in pyrolysis gas chromatography [2]; system 3 is applied in elemental analysis [10]; system 5 in cases when the conversion of the components already separated into products more suitable for detection is preferred [11]; system 7 permits the qualitative identification of the individual fractions leaving the detector by means of chemical reactions [9]; systems 6a, b make the detection of changes in the separated components in the reactor possible [7, 8]; finally, system 8b is applied if the separated fractions are recorded with the help of a chemical detector (see, for example, [16]).

The simple system 8a can be used for the qualitative (and in special cases, also for quantitative) analysis of unknown mixtures. In this system the detection of the fractions separated in the chromatographic column is carried out on the basis of qualitative reactions (group or individual reactions). In system 8a the usual gas chromatographic detectors are not necessary. For the practical use of the system, the apparatus proposed by Casu and Cavalotti [12] can be mentioned.

The simple systems given in Table 1 differ only in the relative position of the reactor. It should be mentioned that the use of other, more complex systems employing several reactors and columns connected either in parallel or in series is also described in the literature. These more-complex systems are considered as combinations of the simple systems.

Thus, for example, for the continuous elemental analysis of organic compounds for the carbon and hydrogen content [13] a system was suggested in which the individual fractions, after separation in the column, entered a reactor filled with copper oxide and iron oxide. The converted product was then separated in a column with acetonylacetone. This complex system can be considered as a combination of systems 0 and 3.

It might be convenient to use for the qualitative identification of pure compounds and compounds which were first separated in a chromatographic column, and also for the group analysis of complex mixtures, specific absorption reactions according to the systems developed by Franc and Michajlova [14]. In this system the identification of the unknown compound is accomplished by means of its retention times in several parallel columns with different stationary phases.

A further possibility is to use several parallel columns which differ in length but which are filled with the same material. Reactors with selective chemical absorbents which permanently retain certain compounds of given classes; e.g., olefins, aromatic hydrocarbons, normal paraffins, etc., are placed after each chromatographic column except one. If the resistances of the parallel gas lines (column, reactor) are approximately equal, then the retention time in each case will be approximately proportional to the column length. If the particular substance analyzed is not absorbed on any of the absorbents which are used, peaks are recorded on the chromatogram; their number will be equal to the number of parallel columns. The peaks are distributed on the chromatogram in sequence and with relative positions depending only on the length of the individual columns used. Therefore, the absence of any particular peak on the chromatogram will indicate that the compound being studied has reacted with that particular chemical absorbent − a useful fact for qualitative characterization.

The same system can also be used for the group analysis of mixtures. In a group analysis, peaks will not disappear completely but − as the result of the formation of nonvolatile compounds in the corresponding reaction − only partially decrease in area.

The apparatus used in analytical reaction gas chromatography is distinguished by its great variety, since in practice different types of chemical reactions which take place under very different conditions are used (pyrolysis, cracking, hydrogenation, dehydrogenation, esterification, oxidation, saponification, etc.).

The interpretation of the results which are obtained is simplified if the chemical reactions satisfy the following requirements: high selectivity, minimum number of formed products, almost complete conversion.

The chemical reactors used can be separated into two groups: flow-type reactors, which are always included in the carrier gas flow, and nonflow-type reactors [10].

In the case of slow reactions, the degree of conversion can be increased by the following method. After the reacting component has entered the reactor, the reactor is bypassed by the carrier gas flow during the time that the reaction takes place, and is then switched back again in order to elute both the unreacted compounds and products formed. If it is impossible to avoid the expansion of the individual zones after the reaction, they can be narrowed by using special techniques (e.g., condensation in cooling traps followed by rapid thermal desorption, narrowing the band with the help of a thermal field, etc.). In some cases (e.g., in carrying out qualitative color reactions [9]) the influence of the diffusion (expansion) of the zones can be neglected.

Flow bubble reactors with an increased capacity for liquid-phase reactions [15] are of particular interest. The application of such reactors significantly expands the circle of reactions which can be used in analytical reaction gas chromatography (in particular it is possible to make wide use of liquid-phase reactions at normal temperatures), and permits a sizable increase in the sample size and thereby improves the sensitivity of the system in the determination of trace impurities.

In selecting the most suitable reactor for a given reaction one must evaluate for each reaction to be performed both the characteristics of that reaction and the effect of the reactor on the chromatographic separation.

LITERATURE CITED

1. A. Keulemans, Gas Chromatography, Reinhold, New York, 1957 [Russian translation: IL, Moscow, 1959].
2. V. G. Berezkin, O. L. Gorshunov, and M. A. Geiderich, Plast. Massy, No. 11:53 (1965).
3. J. E. Hoff and E. D. Feit, Anal. Chem., 36:1002 (1964).
4. N. H. Ray, Analyst, 80:853 (1955).
5. M. W. Anders and G. J. Mannering, Anal. Chem., 34:730 (1962).
6. R. L. Martin, Anal. Chem., 32:336 (1960).
7. W. B. Innes, W. E. Bambrick, and A. J. Andreatch, Anal. Chem., 35:1198 (1963).
8. V. G. Berezkin, A. E. Mysak, and L. S. Polak, in "Gas Chromatography," Tr. II Vses. Konfer., Izd. Nauka, Moscow, 1964, p. 332.
9. J. T. Walsh and C. Merritt, Anal. Chem., 32:1378 (1960).

10.   V. G. Berezkin and V. M. Fateeva, The Chemical Industry
      Abroad, NIITEKhIM, No. 2:65 (1965).
11.   G. E. Green, Nature, 180:295 (1957).
12.   B. Casu and L. Cavalotti, Anal. Chem., 34:1514 (1962).
13.   F. Cacace, R. Cipollini, and G. Perez, Science, 132:1253
      (1960).
14.   J. Franc and S. Michajlova, J. Chromatog., 12:22 (1963).
15.   V. G. Berezkin, A. E. Mysak, and L. S. Polak, Neftekhimiya,
      4:156 (1964).
16.   J. I. Henderson and J. H. Knox, J. Chem. Soc., 2299 (1956).

Chapter IV

The Analysis of Complex Mixtures

Reaction gas chromatography can be most successfully used in analyzing complex mixtures composed of compounds of different classes, each characterized by a different reactivity. In such cases the group reactivities rapidly identify the group composition for the sample and type of compound which corresponds to each chromatographic peak. This simplifies the subsequent identification of the individual peaks significantly.

Analytical reaction chromatography is used less frequently for the analysis of mixtures consisting of compounds of a single type, although in this case even small changes in the structure of the reacting molecules usually result in changes in their reactivity; the successful use of pyrolysis for the identification of hydrocarbon isomers [1] and the diene reaction in the analysis of cis- and trans-diene isomers with conjugated bonds are examples [2].

The use of different types of chemical reactions for the analysis of complex mixtures will be considered below.

HYDROGENATION AND DEHYDROGENATION

The use of these well-studied reactions shows particularly good possibilities in the analysis of complex hydrocarbon mixtures. In all cases the hydrogenation and dehydrogenation products (for olefins – hydrocarbons, for alcohols – hydrocarbons, for naphthenes – aromatic hydrocarbons, etc.) differ greatly from the starting compounds in their chromatographic characteristics (see, for example, [3]); this can be used for the qualitative and quantitative analysis

of the complex mixtures. Therefore many studies have been devoted to the use of hydrogenation and dehydrogenation of different
organic compounds in their identification.

Keulemans and Voge [4] studied in detail the dehydrogenation
of $C_6 - C_8$ naphthenes on a platinum—alumina catalyst (Pt on Al_2O_3).
The dehydrogenation was carried out at 350°C in a hydrogen atmosphere used as the carrier gas. It was shown that the degree of
dehydrogenation increases in going from the C_6 to the C_8 hydrocarbons. Cyclohexane homologs, except for some alkyl-substituted
compounds (of the type 1,1-dimethylcyclohexane), are readily converted into the corresponding aromatic hydrocarbons, and the conversion is complete. Cyclopentane homologs are converted into
the aromatic compounds to a significantly smaller degree (the yield
usually does not exceed a few percent). Rowan [5] recommends
that the dehydrogenation be carried out in the presence of only a
small concentration of hydrogen (3% in the carrier gas) which sup-
presses the dehydrogenation of the cyclopentane homologs but has
no significant effect on the dehydrogenation of the cyclohexane
homologs.

Rowan [5] studied in detail the possibility of using hydrogenation and dehydrogenation reactions in gas chromatography for analytical purposes. He showed that it is not difficult to carry out selective hydrogenation of aromatic hydrocarbons and olefins as a
group reaction. It was shown experimentally that a platinum—
alumina catalyst (1.4% platinum on aluminum oxide) containing a
halogen is best for this purpose. The study of hydrogenation and
dehydrogenation was carried out mainly under the following standard conditions: height of the catalyst layer, 13 cm; reactor diameter, 6 mm; weight of the catalyst, 1.5 g; temperature, 299°C;
flow rate of helium (hydrogen), 60-70 ml/min.

Dehydrogenation is a more complex process, which is often
accompanied by side reactions. Cyclopentane homologs are converted into aromatic hydrocarbons to a much lesser degree than
the cyclohexane homologs.

As general methods for the analysis of complex hydrocarbon
mixtures, Rowan suggested carrying out dehydrogenation of a mixture of isoparaffins and naphthenes for the selective determination
of the concentration of cyclohexanes with the help of the peaks cor-

responding to the aromatic hydrocarbons formed, and also the utilization of hydrogenation at room temperature for the selective
conversion of the olefins to the corresponding paraffins.

Klesment, Rang, and Eizen [6] carried out a detailed study
on the development of a method for the analysis of naphthene hydrocarbons through microcatalytic dehydrogenation and hydrogenation
in conjunction with gas chromatography, and were using this method to study the individual composition of light tar fractions formed
in the distillation of shales. In their method the sample was introduced in a heated copper reactor (diameter 6 mm) filled with 2-3
ml of a platinum catalyst (2-10% platinum on diatomite). The catalyst was prepared from chloroplatinic acid according to the method of Landeberg et al. [7]. The reaction products were separated
in a chromatographic column (600 × 0.6 cm) containing polyethylene
glycol-4000 stationary phase.

It was shown that under the experimental conditions, in a hydrogen flow at 325°C, the yield of the aromatic hydrocarbons
formed as the product of the dehydrogenation of cyclohexane and
cyclohexene homologs was 87-99%. An insignificant amount of de-
alkylation products (benzene, toluene) were formed as a result of
side reactions. Decalin is most difficult to dehydrogenate, and in
this case, along with naphthene, tetralin is also formed. Under
these conditions the isomerization of cyclopentane derivatives into
cyclohexane homologs does not take place, and even the formation
of aromatic hydrocarbons from the cyclopentane hydrocarbons is
not observed. However the process is complicated by hydrocracking reactions.

Hydrogenation was carried out at 200°C in either a current of
hydrogen or helium (due to the hydrogen chemisorbed earlier). The
chromatograms for the original sample and for the hydrogenation
products are shown in Fig. 11.

The given method cannot be recommended for samples containing sulphur compounds or dienes with conjugated bonds, since
these compounds pollute the catalyst.

Lille [8] developed methods for analysis of unsaturated hydrocarbons in shale gasolines with boiling points up to 150°C using
vapor phase hydrogenation in a microreactor placed in front of the
chromatographic column. Reduced nickel (20% on diatomite) was

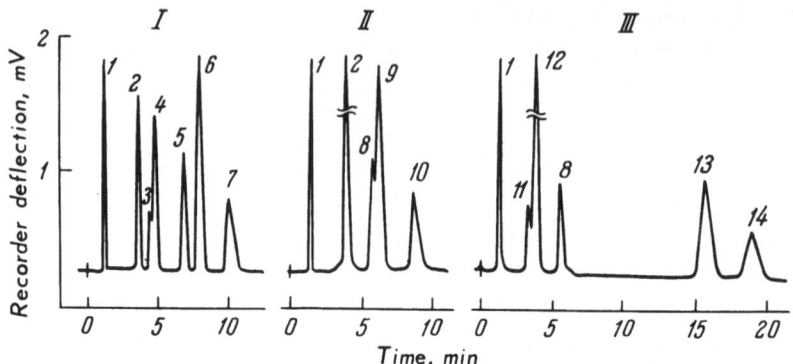

Fig. 11. Chromatograms for the original sample (I), and for the products of hydrogenation at 200°C (II) and dehydrogenation at 325°C (III) [6]. Experimental conditions: carrier gas, hydrogen; flow rate: 60 ml/min; column, 600 × 0.6 cm; column material, 20% polyethyleneglycol-4000 on support material. Peaks: 1) Air; 2) n-octane; 3) octene-1; 4) octene-2; 5) n-propylcyclopentene-1; 6) cyclohexene; 7) 1-isopropylcyclohexene-1; 8) n-propylcyclohexane; 9) ethylcyclohexene; 10) isopropylcyclohexane; 11) product from the dehydrogenation of propylcyclopentane; 12) n-octane; 13) ethylbenzene; 14) isopropylbenzene.

used as the catalyst. The hydrogenation was carried out in a copper reactor (20 × 0.6 cm) at 70°C. It was noticed that during the hydrogenation of disubstituted derivatives of benzene and cyclohexane the trans and cis isomers were formed, and their ratio is determined by the structure of the compounds being hydrogenated and the hydrogenation conditions.

The hydrogenation of unsaturated compounds for identification was also successfully used by Pankov et al. [9].

In studying the composition of complex C_6 hydrocarbon mixtures, Döring and Hauthal [10] used hydrogenation as an important auxiliary method. The hydrogenation was carried out in hydrogen flow (used as the carrier gas), in a small reactor (4 × 0.6 cm), filled with a platinum catalyst (10% platinum on asbestos). The hydrogenation of the olefins and dienes took place quantitatively under chromatographic conditions.

Smith and Ohlson [11] developed a method for the identification of unsaturated hydrocarbons based on their hydrogenation after isolation in a pure form as a result of chromatographic separation. The isolation of the pure substances (or individual fractions) was carried out in a U-shaped trap (total length 15 cm, diameter 0.5 cm) containing an Adams hydrogenation catalyst (1% platinum oxide, layer height 14 cm) cooled in a dry ice bath. After collecting the chromatographic fraction, the trap was removed from the cooling bath and filled with hydrogen to a pressure of 3 atm for 1 min; during this operation, one stopcock of the trap was closed. The trap with closed stopcock was kept in hot water (80-90°C) for 10 min in order to carry out the complete hydrogenation of the fraction collected. After hydrogenation, the trap was connected to the inlet of a gas chromatograph, and the product was transferred into the chromatographic columns for separation by the carrier gas flow. The method was used successfully for straight-chain hydrocarbons and for cyclic compounds with double and triple bonds. In all cases the degree of conversion was closed to 100%. Aromatic hydrocarbons were converted into cyclohexane derivates with a yield in excess of 90%. In order to carry out the partial hydrogenation of dienes, the duration of the reaction was decreased to 5-10 sec. In this study methods were also proposed for the partial hydrogenation of alkines.

Hydrogenation methods were also used successfully in the analysis of compounds of classes other than hydrocarbons. The preliminary hydrogenation in front of the chromatographic column is a rapid and precise method of establishing the carbon skeleton of the compound being analyzed.

Thompson et al. [12-15] developed a series of simple and original methods for the identification of sulfur-, oxygen-, nitrogen-, and halogen-containing compounds by catalytic hydrogenation. The individual isolated fractions (or the pure starting compounds) were hydrogenated on a palladium (0.5% palladium on aluminum oxide) or platinum (5% platinum on porous glass) catalyst and the hydrocarbons which were formed were determined by gas chromatography. The method can be used effectively for the determination of the structure of the carbon skeleton and, in some cases, the position of the heteroatom in the compound being analyzed. Obviously, it is possible to work out a combined, accelerated method in which the hydrogenation and separation would be combined.

Beroza et al. [16, 18] carried out a systematic study of great practical importance concerning the determination of the structure for the carbon skeleton of high-boiling-point compounds by the

method of hydrogenation followed by gas chromatographic analysis
of the products formed. As an example, chromatograms obtained
by combining preliminary hydrogenation and chromatographic
separation of compounds of different types are shown in Fig. 12
[18]. These results were obtained using a chromatograph with a
katharometer. The reactor (24 × 1 cm) was directly connected with
the chromatograph. The neutral hydrogenation catalyst consisted
of 1% palladium deposited on Gas Chrom P. The hydrogenation
was carried out at 280–285°C. The effect of different experimental
parameters (sample size, reactor temperature) on the ratio of
the products formed was also studied. It was shown, specifically,
that at temperatures near 200°C the main reaction is hydrogenation.
In accord with this, methylcyclohexane, toluene, benzyl alcohol,
benzaldehyde, and benzonitrile are converted into methylcyclohex-
anes. On the other hand, at 360°C these compounds will form toluenes.
Beroza, using both thermal-conductivity and flame-ionization detectors in the hydrogenation method, noted that the use of the katharometer permits work with larger samples with isolation of the
reaction products for subsequent study by independent physico-
chemical methods; the use of the flame-ionization detector, however, permits work with very small samples and increases the
efficiency of chromatographic separation.

The problems of identifying the compounds in the hydrogenation method were specially studied by Franc and Kolouskova [19]
who showed that the value of log V^s/V^h (V^s and V^h are the retention volumes of the starting and the hydrogenated compounds, respectively) is determined by the particular group to which the organic substance being hydrogenated belongs. The results for hydrocarbons with various numbers of double bonds as well as for
alcohols, aldehydes, ketones, nitriles, and halogen derivatives,
follow an established rule. In fact, in accordance with the basic
conditions of additivity in gas chromatography [20], the following
equation should be valid:

$$\log V^s/V^h = \Sigma n_{ij}^s\, I_{ij}^h - \Sigma n_{ij}^h\, I_{ij}^h = \Delta n_{fl} I_{fl}^{sh} = \text{const},$$

where n_{ij}^s and n_{ij}^h are the number of structural elements of a given
type in molecules of the starting and the hydrogenated compounds,
respectively; I_{ij}^s and I_{ij}^h are the increments in the logarithm of the
retention volume corresponding to a given structural element in the

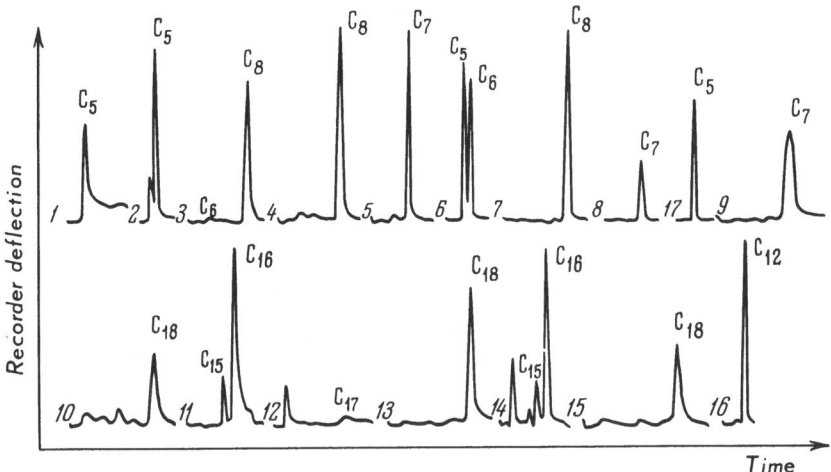

Fig. 12. Chromatogram of the hydrogenation products of the following compounds [18]: 1) hexanoic acid; 2) 3-hexene-1-ol; 3) 2-octanol; 4) nonyl aldehyde; 5) 1-amyl; 6) dihexylether; 7) octyl bromide; 8) 3-heptanone; 9) octadecen-1,12-diol; 10) N, N'-dimethyl-9-octadecenylamine; 11) 1,2-epoxyhexadecane; 12) methyloleate; 13) 9-octadecene-1,12-diacetate; 14) allylhexadecyl ether; 15) octadecylbromide; 16) dodecymercaptan; 17) pentyl disulfide. Experimental conditions: column, 240 x 0.6 cm; column packing, 5% SE-30 silicone gum on a diatomaceous earth support; column temperature, 60°C (peaks 1-8, 17) and 205°C (peaks 9-16); sample volume, 1.0-0.5 µl.

molecule of the starting and hydrogenated compounds, respectively; $\Delta n_{fl} I_{fl}^{sh}$ is the difference in the increments due to the change in the molecules during hydrogenation. For example, in the hydrogenation of $R_i CH_2 Br$:

$$\log V^s / V^h = I_{C-Br} - I_{C-H},$$

and in the hydrogenation of $R_i CH = CHR_i$:

$$\log V^s / V^h = I_{CH=CH-} - I_{-CH_2-CH_2-}.$$

In other words, the value of $\log V^s / V^h$ is determined by the type of functional group which is reduced, and in the first approximation does not depend on the nature of the radical R_i.

In the system of Franc and Kolouskova [19] the sample was a split at the inlet of the reactor into two approximately equal parts, one of which was hydrogenated and the other conducted directly to the inlet. An empty glass capillary tube (diameter ~ 1 mm) was passed through the catalyst layer to divide the sample inside the reactor. Metallic platinum (10% pumice) was used as the catalyst. The hydrogenation temperature was 180°C. The products and the unreacted compounds were separated in a column (length 85 cm) at 68 and 150°C using the 3,5-dinitrobenzoate of the butylether of the triethyleneglycol as the liquid phase.

In this study it was noted that the sulphur and halogen-containing compounds pollute the catalyst and that the hydrogenation of complex esters and of pyridine and its derivatives does not take place.

Franc et al. [21] suggested an interesting method for the identification of alkyl and aryl groups in organic sulphides. This method is based on the preliminary hydrogenation of samples on Raney nickel in a butanol solution (2 ml from a suspension of 20 g Ni in 100 ml n-butanol) at 100-110°C by using a special reactor connected with the gas chromatograph through a valve. After the hydrogenation (~ 15 min), the hydrocarbon products (corresponding to the two radicals of the sulfide hydrogenated) which have been formed are blown out of the reactor and analyzed by gas chromatography. Thus, for example, when analyzing ethylbenzylsulfide only ethane and toluene were found among the hydrogenation products; when analyzing methylpropylsulfide, methane and propane are found, etc. Similar methods may also be developed for the analysis of nonvolatile and compounds of high molecular weight.

The hydrogenation reaction is also used for improving and accelerating separations. Drawert et al. [22, 23] used this reaction to determine small amounts of ethyl alcohol in aqueous solutions and in blood. Raney nickel on kieselguhr (1:10) was used to convert the alcohol into hydrocarbon by hydrogenation. The reaction was carried out at 160-200°C in a hydrogen flow which served simultaneously as the carrier gas. The ethane formed in the reaction was then determined chromatographically. This method can also be used to determine trace amounts of $C_1 - C_{10}$ alcohols in aqueous solutions.

Klesment developed an interesting method which makes it possible to record the hydrogen concentration of carrier gas (a mixture of nitrogen and hydrogen) which was dissolved during dehydrogenation or consumed in hydrogenation [76].

All the studies discussed above were carried out using the usual packed analytical columns. However, chemical reactions can also be utilized in capillary chromatography, particularly when complex mixtures are being studied (and therefore it might be possible that if packed columns are used, the peaks of the reaction products would be overlapped) or when the chromatogram of the reaction products is complex. Struppe [24] used reaction gas chromatography in conjunction with capillary chromatography. An aluminum capillary tube (600 × 0.03 cm) the inner walls of which were coated with a thin platinum layer served as the reactor. The catalyst was deposited on the inner wall of the capillary reactor by the usual method of deposition with a stationary liquid phase in a capillary column, i.e., the capillary was filled with an ether solution of chloroplatinic acid which, in the course of 15 min, moved from one end of the tube to the other. The reactor was then heated at 150°C in a hydrogen flow, thereby reducing the chloroplatinic acid to platinum. The hydrogenation process was carried out in hydrogen flow at 125°C. The method was checked by analyzing synthetic mixtures of hydrocarbons with boiling points up to 85°C. It was shown that mono-, di-, and cycloolefins add hydrogen rapidly at the double bond, in which case the carbon structure of the aromatic, naphthene, and cyclic hydrocarbons does not change during hydrogenation under the given conditions.

In conclusion it must be pointed out that at the present time simple and sufficiently effective methods for hydrogenation and dehydrogenation have been developed only for the analysis of a single compound; a satisfactory hydrogenation and dehydrogenation method for the direct identification of unsaturated compounds in complex mixtures during the chromatographic analysis has not yet been realized.

ESTER PREPARATION AND HYDROLYSIS

The straight gas chromatographic analysis of organic acids is complicated by the fact that they are polar compounds. Therefore acids (particularly the higher acids) are usually analyzed in the form of their esters; in this way, the efficiency of the separation is increased, there are fewer asymmetric peaks in the chromatogram, and the separation temperature is decreased. The esterification of the fatty acids is carried out either independently of the chromatographic analysis or in the chromatograph in front of the column [25].

Ralls [26] developed a method for the rapid preparation of the ethylesters of the $C_1 - C_7$ acids in a capillary connected to the inlet of the chromatograph. The dry potassium salts of the aliphatic acids were carefully mixed with an equal amount, by weight, of potassium ethyl sulfate. The sample (4-5 mg) was placed in a capillary of borosilicate glass closed at one end (length 11.5 cm; diameter: outside, 1.5 mm; inside, 2 mm) which was connected through a silicone rubber gasket with the sample-injection chamber of the chromatograph. Upon heating the lower part of the capillary in a silicon bath at 300°C for 10 sec, the ethylesters of the acids are formed, which volatilize and enter the chromatographic column. The described method, however, is not completely quantitative. Stephens and Teszler [27] described a refined method for connecting the capillary with the sample-injection chamber which can also be used to carry out other reactions.

Hunter [28] suggested a simplified version of the method for fast double-decomposition reactions in which solutions of mixtures of the salts of the acids being analyzed and potassium ethylsulfate are used initially. In a test tube, 0.1 ml solution of the sodium or potassium salt of $C_1 - C_{18}$ aliphatic acids being studied is mixed with 0.1 ml potassium ethylsulfate solution and 0.01 ml red ink. A hypodermic needle filled with Celite is placed in this solution and after adsorption the needle is dried at 105°C for 25 min. Then, by connecting the needle with a source of argon, it is introduced in the usual way through a rubber septum into the evaporation chamber heated to 275°C, at which temperature the double-decomposition reaction takes place rapidly. Twenty-five seconds after the introduction of the needle into the heated chamber, the ester formed is blown out by argon flow into the column. The method was used successfully for the identification of aliphatic acids in fermented liquids. The questions of quantitative analysis were not studied in this work.

The methylesters of organic acids can also be obtained by the pyrolysis of their tetramethylammonium salts carried out in the heated injection block of the instrument [29]. It was recommended that the tetramethylammonium salts of the acids be prepared either by the titration of a methanol solution of the acid with an $N(CH_3)_4OH$ solution or via ion exchange [30]. Pyrolysis of the tetramethylammonium salts of the acids is carried out in an evaporator at 330-365°C. The yield of the esters of acetic, butyric, valeric, caprylic, lauric, myristic, palmitic, benzoic, cinnamic, levulinic, glycolic, and lactic acids is 86-99%. The esters of dibasic acids (oxalic, lactic, malic, and citric) are not formed under these conditions. It was shown that the yield of the esters in the pyrolysis of the tetrammonium salts decreases for samples smaller than 0.05 mg. The method can be recommended for qualitative and semiqualitative analyses (when a high degree of accuracy is not necessary).

Drawert et al. [22, 31, 32] developed a method for the analysis of alcohols and dilute aqueous solutions in the form of the esters of nitrous acid. The esters were formed in the carrier gas flow in a reactor (length, 8-12 cm; diameter, 10 mm) which was installed upstream of the chromatographic column. The reactor was filled with Sterchamol* impregnated with sodium nitrite (1:1). The esters of the nitrous acid were formed at 160-200°C. The process of ester formation is improved and will be quantitative if the sample introduced is first acidified with tartaric or oxalic acid. The reactor can also be filled with a mixture of oxalic acid and firebrick and the solution of sodium nitrite added along with the sample of alcohols to be analyzed.

Anders and Mannering [33] suggested a method for the preparation of the esters directly in the column. A few seconds (no more than 5 sec) after the sample is introduced, acetic or propionic anhydride is injected into the column; it will react in the column with the compounds containing $-OH$, $-NH_2$, or $=NH$ groups to form the corresponding derivatives.

The products formed are characterized by different retention times as compared with the original substances, resulting in the appearance of new peaks in the chromatogram. These peaks are used for the identification of the compounds being analyzed. The degree of conversion of the original sample components (alkaloids and steroids) varies over broad limits, increasing with an increase in the amount of the acylating agent (the anhydride). In the study it was shown that up to 50 μl anhydride can be used without

_____

*Sterchamol is a German diatomaceous earth-type support.

significant tailing of the chromatographic peaks. Propionic anhydride is the stronger acylating agent. The size of the sample
varied within 0.5-8 μl of 0.5% solution. The separation and formation of the derivatives was carried out in a column containing SE-
30 methylsilicone polymer stationary phase (2% on GasChrom S) at
different temperatures. The authors also showed that it was possible to form trifluoroacetates and trimethylsilicyl esters directly
in the column by using an analogous method.

In analytical reaction gas chromatography, hydrolysis reactions are also widely used both for the identification of compounds
and for their quantitative analysis.

The preliminary hydrolysis before chromatographic separation is a suitable way of analyzing unstable and reactive compounds
which form stable products in the process.

Thus, for example, Molinary et al. [34] developed a method
for the analysis of readily hydrolyzable metalloorganic compounds
which includes the preliminary hydrolysis of the sample in a reactor in the presence of phosphoric acid. The method was checked
by analysis of an ether solution of magnesium ethyliodide. After
the sample ($\sim$ 50 ml) was hydrolyzed, the light hydrocarbons
formed were determined on a column containing silica gel. The
mean error was $\pm$ 3.3%.

A simple method for the analysis of partially decomposed
Grignard reagents, making possible the determination of their concentration in a solution, was described by Guild et al. [35]. In order to ensure a minimum contact of the Grignard solution with the
atmosphere, a simple attachment with a closed container ($\sim$ 10 ml)
is used for the sample and it is connected by means of a silicone
septum directly with the inlet of the column. Several milliliters of
the Grignard reagent (n-tert-butylphenylmagnesium bromide) and
$\sim$ 1 ml of n-nonane, which is used as the internal standard (the
ratio of the reagent solution and the internal standard are determined gravimetrically), are placed in the container and the attachment is connected to the inlet of the chromatographic column.
An aliquot sample of the mixture of the Grignard reagent and the
internal standard is taken out by a microsyringe, and about 0.065
ml is introduced into a propyleneglycol column. The amount of
reagent which has decomposed during standing is determined from
the peak of the eluted tert-butylbenzene. After the complete de-

composition of the Grignard reagent present in the container with methanol, 0.005 ml of the sample is again injected into the column, determining in this case the total amount of the reagent in solution. The results of the analysis of the Grignard reagent by the chromatographic method are in good agreement with those obtained by titration.

This method for the analysis of the catalyst for polyethylene synthesis (triethylaluminum) is also based on preliminary hydrolysis [36]. The hydrolysis was carried out in a reactor filled with 30% lauric acid on Sil-O-Cel (50-80 mesh). The hydrolysis products were determined chromatographically: ethane and butane on a dibutylphthalate column (carrier gas, hydrogen), hydrogen and ethane on silica gel column (carrier gas, nitrogen). The analytical results permit calculation of the concentration of the active material in the product being studied. It should be pointed out that 3-methylpentane, 3-methylheptane, n-hexane, and n-octane were found among the hydrolysis products. The method can also be used for the analysis of $Al(CH_3)_3$, $Al(C_3H_7)_3$, and $Al(C_4H_9)_3$.

The hydrolysis products formed make it possible, in a number of cases, to identify and quantitatively determine some of the compounds which are present in the mixture being hydrolyzed. Thus, for example, a method was developed [37] for the analysis of $BHCl_2$ in a mixture with BCl_3 and HCl, based on the preliminary hydrolysis of the compounds being analyzed in a column with a 5-Å molecular sieve and on the subsequent removal of hydrogen from the other products. If hydrogen is present in the starting mixture, it must be removed before the hydrolysis of the sample; this can be accomplished on another molecular sieve column or, for example, by using cooled traps.

The direct chromatographic analysis of mixed high boiling esters with high boiling points is a complicated task. In this connection, Janak, Novak, and Sulovsky [38] suggested using consecutive chromatographic separation, saponification, and chromatographic analysis of the resulting products for the identification of complex esters. A reactor filled with Sterchamol with 40% potassium hydroxide deposited on its surface was connected to the outlet of a laboratory chromatograph with a katharometer. The effluent from the detector with the separated zones of the malonic acid ester fractions and steam were continuously fed into the reactor

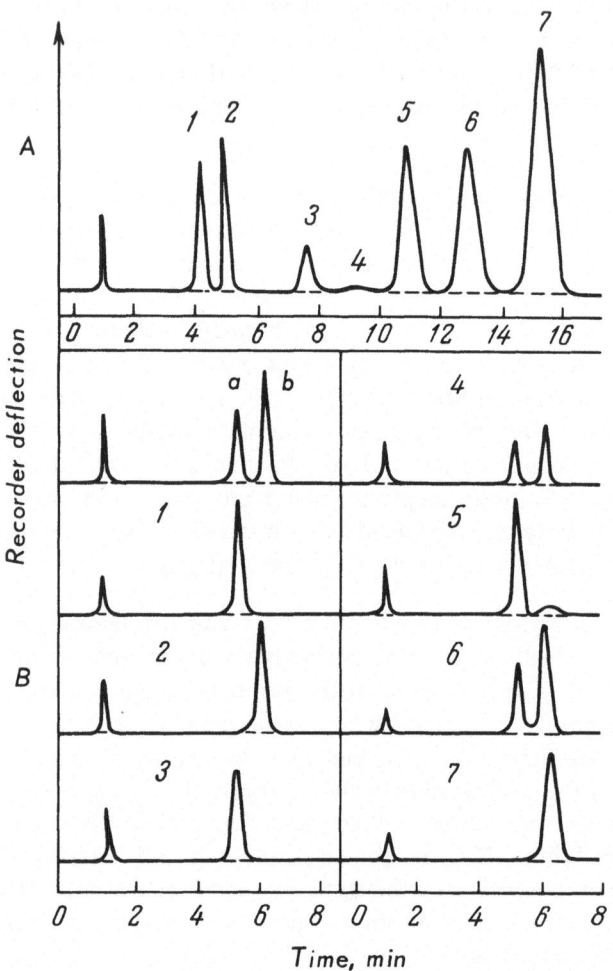

Fig. 13. Identification of esters from the saponification products [38]. A) Chromatogram of the mixture of malonic esters;
B) chromatogram of the alcohols in an artificial mixture (a —
methanol, b —ethanol) and a sample of the saponified fractions
corresponding to the peaks 1-7 in chromatogram A. 1) Methylester of phenylacetic acid;   2) ethyl ester of phenylacetic acid;
3) dimethyl ester of   phenylmalonic acid; 4) methylester of
phenylmalonic acid; 5) dimethyl ester of ethylphenylmalonic
acid + traces of diethyl ester of phenylmalonic acid; 6) methylethyl ester of ethylphenylmalonic acid; 7) diethyl ester of ethylphenylmalonic acid.

inlet. After saponification of the high-boiling esters, the effluent from the reactor containing the corresponding alcohols formed in the hydrolysis of the complex esters was sampled and the sample analyzed in a chromatograph with a flame ionization detector in order to determine the alcohols formed.

Figure 13 shows the chromatograms of the saponification products for different esters of malonic acid. Apiezon L (at 220 °C) was used as the stationary phase for the separation of the original mixture of the malonic esters, and triethanolamine (at 60 °C) for the separation of the alcohols.

The hydrolysis method can also be used for the determination of water. Hager and Baker [39] suggested that, for the determination of 0.03-0.5% water in organic liquids, the sample be treated with 2,2-dimethoxypropane in the presence of methylsulfonic acid. In an acid medium 2,2-dimethoxypropane reacts with water and forms acetone and methyl alcohol [40]. The reaction products were separated at 56-60°C on a column (305 cm) containing 10% QF-1 fluorosilicone oil on Gas Chrom P (100-140 mesh), with a helium flow rate of 70-80 ml/min. The water concentration can be calculated then from the amount of products formed or from the decrease of the 2,2-dimethoxypropane peak.

The methods of reaction gas chromatography have found wide use in the analysis of aqueous solutions. The usual chromatographic method becomes complicated in this case for the following reasons:

1) The water peak is usually broad, asymmetric, and often superimposed over the peaks of other components present in the sample;

2) the direct determination of traces of water is usually not possible because of the low sensitivity of the most widely used detectors (thermal conductivity and flame ionization) to water and the broadness of the water peak. Therefore it is usually necessary to carry out the preliminary, selective absorption of water (to dry the solutions to be analyzed) and to convert the water into compounds which are eluted from the column before the other sample components can be recorded with a flame ionization detector.

Thus, for example, Kung et al. [41] developed a method based on the conversion of water into acetylene in a special reactor con-

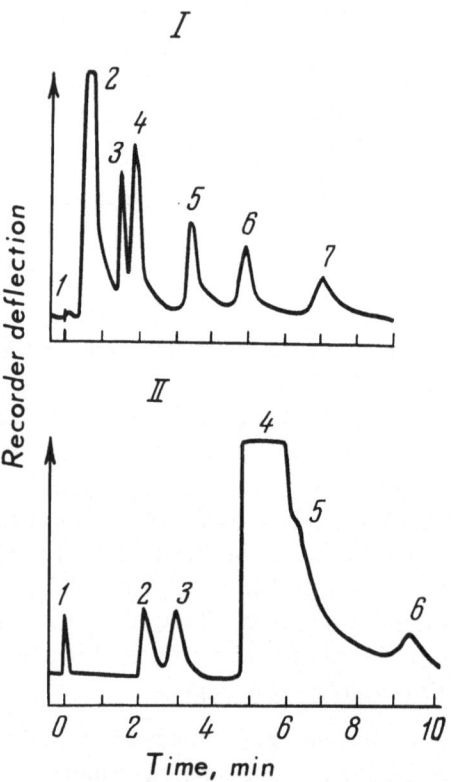

Fig. 14. Chromatograms of alcohol mixtures
containing 90% water: (I) by using a calcium
carbide reactor and (II) without the reactor [41].
I: 1) Air; 2) acetylene; 3) methanol; 4) ethanol;
5) 1-propanol; 6) 2-methyl-1-propanol; 7) 1-bu-
tanol. II: 1) Air; 2) methanol; 3) ethanol; 4)
water; 5) 1-propanol; 6) 2-methyl-1-propanol.

taining calcium carbide placed upstream from the chromatographic
column. The conversion of the water into acetylene was carried
out at 220°C in a pyrex reactor (30 × 1.8 cm) filled with a mixture
of calcium carbide (30 mesh) and glass beads (diameter 0.5 mm)
in a 1:2 ratio. The method was used for the analysis of aqueous
solutions of aldehydes, esters, and alcohols. Organic acids are
retained by the calcium carbide, and therefore this method cannot
be used if they are present in the sample.

Figure 14 shows two chromatograms illustrating the analysis of aqueous solutions of alcohols with a reactor (I) and without a reactor (II). The chromatographic separation was carried out at 74°C on a column (250 × 0.7 cm) containing a polar phase, Ucon Oil 50HB-200. The chromatograms illustrate that the use of a calcium carbide reactor permits, for example, the analysis of alcohols like 2-methyl-1-propanol and 1-butanol which would not be possible without the elimination of the water peak.

The conversion method was also used successfully to determine traces of water, since the narrow peak for acetylene is recorded with a high sensitivity by the katharometer; the use of a flame ionization detector further increases the sensitivity of the system [42-43].

Methods are also described in the literature where the conversion of water by means of calcium carbide [22] is used, specifically, for the analysis of the deuterium concentration of water [44].

It must be pointed out that the conversion of water into acetylene or hydrogen is a heterogeneous reaction, which is complicated by the formation of a layer of the solid product on the surface of the reacting particles. This fact is always a potential reason for experimental errors when using calcium hydride or calcium carbide for the conversion.

The method suggested by Berezkin et al. [45-46] for the determination of traces of water in hydrocarbons, based on the chromatographic determination of the amount of hydrogen formed in the reaction of water (dissolved in the liquid sample) with diethylene glycol dimethylether solution of sodium aluminum hydride, is free of these possible error sources.

In conclusion it must be emphasized that the use of the esterification and hydrolysis reactions permits the solution of such important analytical problems as the qualitative identification of acids and mixed esters, and makes it possible to carry out the analysis of reactive, readily hydrolyzable compounds, as well as of water and aqueous solutions. In our opinion, the further development of these methods in the field of identification and chromatographic analysis of metalloorganic compounds has particularly good prospects.

PYROLYSIS OF VOLATILE SUBSTANCES*

In gas chromatography the reactions resulting in the formation of a number of products instead of one compound (for example,
pyrolysis) are more and more widely used. The chromatogram
for these products forms a characteristic spectrum which can be
used successfully to identify the starting compound. Naturally, in
the case of the pyrolytic method of identification, the whole sample mixture should not be subjected to pyrolysis, but rather a single,
chromatographically separated component (or pure compound),
since the interpretation of the results in the case of the simultaneous pyrolytic breakdown of the entire mixture is very complicated.
Unfortunately, up to the present time the additivity of the observed
picture of the pyrolysis with respect to the mixture of pyrolyzed
compounds has not been clarified. Obviously, in the range of dilute solutions (small samples) the conditions of additivity should
be fulfilled.

Pyrolysis as an accessory medium for identifying volatile
organic compounds was first used by Dhont [47, 48]. The compounds to be analyzed were introduced into the carrier gas flow in
a short quartz reactor filled with Chromosorb. The pyrolysis products were separated on a chromatographic column containing a
silicone oil stationary phase. As the result of studying the pyrolysis products from a large number (about 100) of high-boiling organic compounds, Dhont showed that the benzene rings are not decomposed during pyrolysis and that aromatic compounds are present in the decomposition products. Thus, for example, benzene,
toluene, and carbon dioxide are the main products in the pyrolysis
of benzylalcohol, benzylbenzoate, and benzylphenylacetate.

It was shown, using as an example the study of the pyrolysis
of different alcohols and complex isoamylesters of a number of
organic acids at 600°C, that the pyrolysis products give characteristic chromatograms from which the starting compounds can be
identified. The method gives reproducible results, and is independent of the size of sample pyrolyzed.

Keulemans and Perry [1] demonstrated the great possibilities
of the pyrolytic method for the identification of paraffin hydrocar-

---

*In this section, only the pyrolysis of volatile compounds will be studied. The pyroly-
sis of polymers and other nonvolatile compounds will be described in Chapter VI.

bons. The pyrolysis was carried out in an empty quartz tube at 500°C. Using the analysis of the hexane isomers (2,2-dimethylbutane and 2,3-dimethylbutane) as an example, they established the correlation between the observed products and the possible cleavage of the molecule with respect to different C–C bonds. Keulemans and Cramers [49] further refined the pyrolytic method by using an inert, gold reactor (length, 1 m; diameter, 1 mm) and efficient columns for the separation of the pyrolysis products. As shown by them, cis- and trans-isomers give similar qualitative pictures of the pyrolysis products, but the degree of conversion is different. In some cases pyrolysis gives more valuable results than mass spectrometry. Thus, for example, 2-methyl-pentane-2 and 4-methyl-cis-pentene-2 give similar mass spectra, but sharply different chromatographic spectra, of the pyrolysis products. The pyrolysis method is more simple but enables approximately the same analytical information to be obtained as mass spectrometry. The reproducibility of both methods is almost the same.

It is also convenient to use preliminary pyrolysis for the analysis of mixtures which contain thermally unstable compounds [50]. Thus, for example, in the injection block heated to 310°C, a rapid decomposition of the peroxide takes place in the analysis of solutions of di-tert-butylperoxide. On the basis of the pyrolysis products, some definite conclusions can be drawn concerning the structure of the peroxide, and its concentration in the solution can be determined. In the study cited above, a linear relationship was found between the peroxide concentration in the sample and the area of the peaks corresponding to the decomposition products.

The pyrolytic method of identifying organic compounds is one of the most prospective methods in analytical reaction gas chromatography, although at the present time its possibilities are not yet completely clear. Since reproducible results can only be attained under standard conditions, its broad development and the establishment of relationships between the structure and pyrolysis products are held back because of the absence of a standard method and standard apparatus.

DEHYDRATION AND DECARBOXYLATION

The methods for the dehydration of alcohols and the decarboxylation of monobasic organic acids were used by Drawert and Kupfer

[31] for the analysis of different mixtures. The sample to be analyzed first entered a stainless steel reactor (10–15 × 0.6–1.0 cm)
filled with Sterchamol treated with phosphoric acid (weight ratio
2:1). At 250–300°C, monobasic organic acids are decarboxylated,
forming hydrocarbons in which the number of carbon atoms in the
molecule is one less than that in the starting acid. The alcohols
are converted into the corresponding olefins, in which case two
olefins are formed in the dehydration of secondary alcohols. The
water present in the sample and formed in the dehydration is removed in a drying tube containing calcium hydride which is placed
between the reactor and the chromatographic column.

A multistage reactor was used for the analysis of dilute
aqueous solutions of glycerine and alcohol [32] in which, as the result of consecutive dehydration, hydrogenation, and dehydration reactions, a mixture of propylene and ethane were formed which can
be analyzed easily by gas chromatography.

REACTIONS FOR THE FORMATION
OF NONVOLATILE COMPOUNDS

In this chapter reactions of different types are studied which
result in the formation of almost nonvolatile products. The selective removal of compounds of given classes from the mixture being analyzed is an effective method of identification and quantitative analysis of the components of complex mixtures. The comparison of two chromatograms, obtained in the chromatographic
separation under identical conditions which differ only in the fact
that in one case a reactor, selectively retaining compounds of a
given type, is placed between the chromatographic column and the
detector, and in the second, the usual analysis without the reactor
is carried out, makes it possible to carry out the group identification of the sample components and to obtain quantitative data on
their concentration in the original mixture.

The usual variation of the method of selective removal of
certain components has shortcomings associated with the necessity
of obtaining two chromatograms and with the low quantitative accuracy when the concentration of the removed component is small
in the general, unresolved chromatographic peak.

These methods have been developed in greatest detail for the analysis of hydrocarbon mixtures of different classes.

Martin [51] was the first to use a chemical absorber in gas chromatography for the analysis of unsaturated hydrocarbons in a mixture with paraffins. A reactor (2.0 × 0.4 cm) containing silica gel impregnated with concentrated sulfuric acid was used for the selective absorption of the $C_3 - C_6$ olefins. The absorbent was prepared by mixing 3 parts (by weight) of concentrated sulfuric acid and 2 parts of silica gel (60-200 mesh fraction). Silica gel is a good support for sulfuric acid; it remains porous even after the absorption of more than its own weight of the acid. The use of other supports (cement, firebrick) was found to be ineffective. The prepared absorbent must be stored in closed containers since olefins are not quantitatively absorbed if the water content of the sulfuric acid is about 12%.

Concentrated sulfuric acid on silica gel at 20-50°C completely absorbs monoolefins, diolefins, cycloolefins, and acetylene hydrocarbons, except for acetylene and ethylene.

Sulfuric acid saturated with silver sulfate must be used for the absorption of acetylene and ethylene. The absorbent for the olefins did not react with saturated hydrocarbons.

The chromatographic separation of the hydrocarbons was carried out on a column (1050 × 0.6 cm) containing isoquinoline on firebrick in a ratio of 1:9.

The analysis of a complex mixture can be carried out by two methods. In the first case the mixture is analyzed twice, with and without the absorption reactor; in the second case a chromatograph is used with two independent detectors, the first placed after the chromatographic column (recording all components) and the second placed after the reactor (recording the saturated hydrocarbons only).

Figure 15 shows two chromatograms, one with and one without an absorber. In the analysis using the sulfuric acid reactor, all unsaturated compounds are absorbed and the detector records only the saturated hydrocarbons (chromatogram B). The chromatogram of the analysis using the reactor can be obtained from the usual chromatogram A by subtracting the components which react. Therefore this method is also called the "subtraction method" [51].

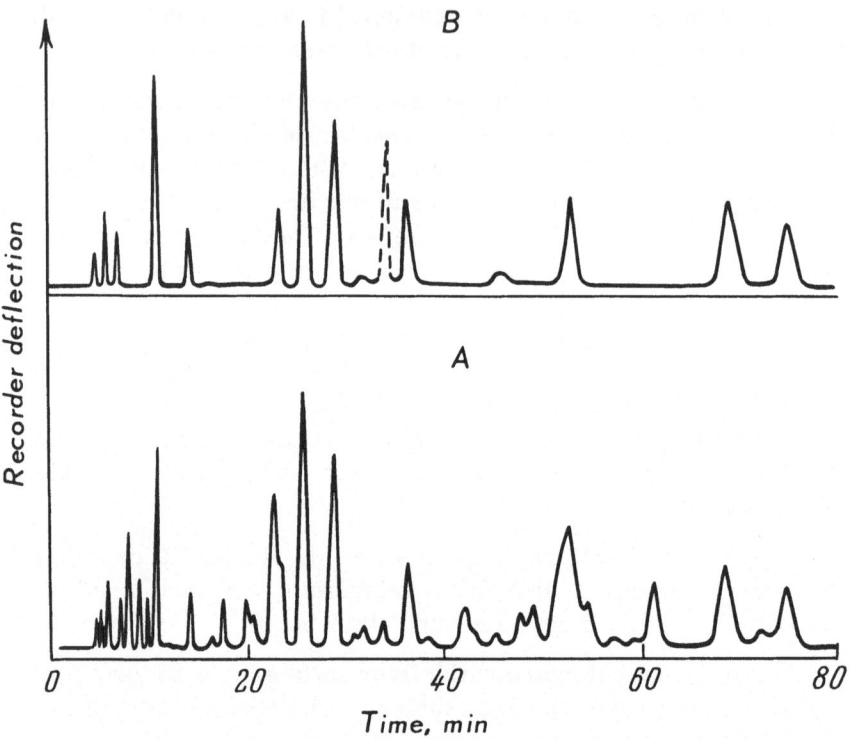

Fig. 15. Chromatogram of hydrocarbons obtained from catalytic cracking of petroleum (C_3–C_6) [51]. A) Analysis without the reactor; B) analysis with the reactor [con-
centrated H_2SO_4 on silica gel]. Analytical conditions: column material, 10% isoquinoline on firebrick; temperature, 25°C; column length, 10.5 m; carrier gas (nitrogen) flow
rate, 60 ml/min; sample volume, 5 µl.

In the analysis of the products from the catalytic cracking of
petroleum, the deviation of the results from the mean values in six
experiments for completely separated compounds did not exceed
1.9%; for olefins, determined by the subtraction method, this value varied within the limits of 2-9%.

Rowan [5] developed a method for the chemical absorption of
certain components of a complex mixture using the following as
group absorbents: concentrated sulfuric acid (absorption of olefins
and aromatic hydrocarbons), mercuric perchlorate (absorption of
olefins), and molecular sieves (adsorption of n-paraffins). The
suggested method of operation was to pass the sample through a
chromatographic column, detector, reactor, and cool trap. The

condensed products were then analyzed again on the column to determine the changes which have taken place in the original sample after a certain type of reaction has been carried out. Rowan also used hydrogenation of the olefins and aromatic hydrocarbons and dehydrogenation of the naphthenes along with the other removal reactions.

The analysis was carried out in a slightly modified standard chromatograph (additional four-way valves, connecting lines, traps, and reactors). The reactors (absorbers) were replaceable. The chromatographic separation was carried out at 125°C, on a 2-m column containing dodecylphthalate stationary phase, using hydrogen or helium as the carrier gas. The sample size varied from 0.01 to 0.05 ml. The cool trap was a U-shaped copper tube (diameter, 3 mm) and it was recommended that a small part (5 cm) of it be filled with a column packing (35% DC-200 silicone on firebrick). The experiments showed the effectiveness of such a trap. Liquid nitrogen was used for cooling, although in many cases dry ice can also be used.

The selective absorption of the olefins and aromatic hydrocarbons was carried out in an absorber (diameter, 6 mm) which contained 2 ml concentrated sulfuric acid deposited on fiberglass (length of the packing, 43 cm). A 4-$\mathring{A}$ molecular sieve and a small amount of Ascarite were packed at the outlet of the absorber. This packing adsorbs methane, ethane, acetylene, propylene, and water, but does not adsorb propane and the other higher-molecular-weight compounds. At 54°C this reactor – absorber absorbs olefins, toluene, and the higher aromatic hydrocarbons; benzene is not completely absorbed.

As pointed out by Rowan [5], the removal of the olefins with sulfuric acid has one shortcoming: 1-2 h after the olefins have passed through the absorber, small quantities of inseparable light hydrocarbons (3-6% of the olefin content) will be continuously eluted from the sample which would be analyzed at that time.

The selective removal of olefins was carried out in an absorber containing mercuric perchlorate prepared, according to the recommendation of Coulson [52], by treating firebrick (40-60 mesh) with an aqueous solution of 1 M mercuric perchlorate and 2 M perchloric acid (in 1:1 gravimetric ratio) followed by drying at 110°C.

No decrease in activity was observed when storing the prepared
absorbent in a closed container. The reactor for the removal of
olefins was a 25-cm-long copper tube (diameter, 6 mm) filled with
the mercuric perchlorate packing; it was kept at 82° or 100°C. A
12.5-cm-long tube filled with a 4-Å molecular sieve and connected
to the outlet of the mercuric perchlorate reactor was used to absorb the water evolved in this reactor. The method of selective
absorption of unsaturated compounds was used successfully for the
analysis of gasoline.

Martin [53] suggested a combined method for the determination of aromatic, olefinic, and saturated hydrocarbons in gasolines,
based on group chromatographic separation and chemical absorption in a reactor containing mercuric perchlorate. At first the
sample is analyzed on a column containing β,β'-thiodipropionitrile
as the stationary phase. The saturated and olefinic hydrocarbons
are eluted first, as a large asymmetric peak, while aromatic hydrocarbons emerge later. During this analysis, the absorber containing mercuric perchlorate and a trap cooled with liquid nitrogen are
connected in series after the detector. The olefins are absorbed
in the reactor and the saturates are collected in the liquid nitrogen
trap. Before the lowest aromatic hydrocarbon emerges from the
column, the parts of the system are rearranged so that the carrier
gas flows from absorber to trap to chromatographic column (entering at its former end) to detector. This backflushes the aromatics
from the column as one peak. Finally, the trap is warmed and the
saturates return to the column and are detected alone. The olefinic hydrocarbons are determined from the difference of the first
"saturated–unsaturated" and the last "saturated" peaks. The merit
of this method is that it is possible to carry out group analysis
from a single sample.

Innes et al. [54] carried out a detailed study of different
chemical absorbers. The appropriate chemical reagent was coated
on a diatomite support (80-100 mesh) in a ratio of 1 ml reagent solution to 1 g diatomite. Reactors 10 cm long with 6.25-mm diameters were used for the selective absorption; the sample size was
5 ml, concentration 0.5%, and flow rate of the carrier gas 20-50
ml/min. The results of testing various reagents are given in
Table 2.

The data given in Table 2 show that a $Ag_2SO_4 - H_2SO_4$ solution (SSSA)* quantitatively absorbs the acetylenic, olefinic, and
_____
*SSSA is an arbitrary designation for the solution of silver sulfate in sulfuric acid.

TABLE 2. Absorption of Different Types of Hydrocarbons by Chemical Absorbents

Hydrocarbon	Absorption, %					
	$20\%\ HgSO_4$ + $20\%\ H_2SO_4$	Saturated solution of $Hg(CH_3COO)_2$	$4\%\ AgSO_4$ + $95\%\ H_2SO_4$	$95\%\ H_2SO_4$	$80\%\ H_2SO_4$	$60\%\ H_2SO_4$
Methane	0	0	0	0	0	0
Ethane	0	0	0	0	0	0
Propane	0	0	0	0	0	0
n-Butane	0	0	0	0	0	0
n-Hexane	0	0	0	0	0	0
n-Octane	0	0	0	0	0	0
Cyclohexane	0	0	8	31	0	0
Ethylene	100	94	100	11	6	0
Propylene	100	100	100	100	67	0
Isobutylene	100	100	100	100	100	100
Pentene-2	100	67	100	100	88	0
Heptene-2	100	60	100	100	85	0
4-Methylcyclohexene	100	70	100	100	70	0
Benzene	5	46	100	94	32	13
Toluene *	0	22	100	100	33	0
p-Xylene *	0	0	100	100	0	0
Acetylene *	100	100	100	10	11	0

* Accuracy ~ 10%.

aromatic hydrocarbons while the solution $HgSO_4 - H_2SO_4$ (MSSA)*
absorbs only the olefinic and acetylenic hydrocarbons. Concentrated sulfuric acid only removes the olefinic (starting with propylene) and aromatic hydrocarbons; ethylene and acetylene are
only partially absorbed.

Innes et al. [54] used an original, differential system to record the results of the chromatographic separation and chemical
absorption which consisted of two flame ionization detectors and
parallel absorbers; one of the detectors recorded the concentration of the hydrocarbons after the first absorber and the second
recorded the concentration after the second absorber. By synchro-
nizing the operation of both detectors with respect to time and by
obtaining a differential signal from the two detectors (electrical
subtraction) it is possible to get an electrical signal which is proportional to the concentration of the hydrocarbons which are being
absorbed (olefinic, acetylenic, etc.). Some variations of using absorbers and detectors are shown in Table 3. The method was used
for the analysis of olefinic and paraffinic hydrocarbons in exhaust
gases and for the analysis of gasolines.

A simpler method was suggested by Berezkin et al. [55] for
the differential recording of selectively absorbed components which
does not require a special synchronization with respect to time.
In this case a selective absorber is placed between the chambers
of the katharometer (or other detector). The analyzer must fulfill
the following requirements: it must record very small, rapidly
changing concentrations of the individual substances in the chromatographic zone on a background of the carrier gas and unabsorbed
compounds; the dead volume of the absorbent and detector chamber
must be small as compared to the carrier gas volume in which the
components are distributed (chromatographic zone); the detector
must have a wide dynamic range; the absorbent must have a
large capacity and absorb compounds of a given class rapidly; the
sensitivity of both sides of the katharometer should be almost the
same. The method of operation was simplified by using the technique of step chromatography.

Albert [57] developed a gas chromatographic method for the
determination of hydrocarbon types (i.e., aromatic, unsaturated,

---

*MSSA is the arbitrary designation for the solution of mercury sulfate in sulfuric acid.

TABLE 3. Variations of Using Different Absorbers

Absorbent			Type of hydrocarbon analyzed	
First absorber	Second absorber	First absorber	Second absorber	Differential signal
SSSA*	MSSA†	Paraffins	Paraffins and aromatic hydrocarbons	Aromatic hydrocarbons
SSSA*	Glycerol	Paraffins	All hydrocarbons	Aromatic, acetylenic and olefinic hydrocarbons
MSSA†	Glycerol	Paraffins and aromatic hydrocarbons	All hydrocarbons	Olefinic and acetylenic hydrocarbons

*Solution of silver sulfate in sulfuric acid.
†Solution of mercury sulfate in sulfuric acid.

n-paraffins, and isoparaffins) in mixtures of the $C_5 - C_{11}$ hydrocarbons. The method is based on the use of a selective triple system: a stationary liquid phase [N,N'-bis-(2-cyanoethyl) formamide] from which the aromatic hydrocarbons are eluted after the other compounds, molecular sieves which selectively retain the n-paraffins, and an absorber containing mercuric perchlorate which absorbs the unsaturated compounds. The absorber is filled to a height of 7.6 cm with mercuric perchlorate on Chromosorb, to 5.1 cm with anhydrous magnesium perchlorate, to 5.1 cm with Ascarite, and to 2.5 cm with anhydrous magnesium perchlorate. The analysis is carried out on a special chromatographic apparatus which consists of a chromatographic column, an absorber, and a trap for repeated chromatographic analysis of some of the groups of hydrocarbon (isoparaffins, n-paraffins). The n-paraffins adsorbed by molecular sieves are desorbed into a trap by heating the sieve to 390–400°C for 15 min. The total analysis time is 1.5 h. The method was used for the analysis of gasolines.

The use of the method of forming nonvolatile compounds in gas chromatographic analysis of hydrocarbons is also described in the literature [9, 58].

The reactions for forming nonvolatile compounds were also used for the selective removal of alcohols from a mixture of organic compounds. Ikeda et al. [159] used a reactor (15 × 0.6 cm) filled with 3% boric acid on Chromosorb P for the removal of alcohols by the formation of nonvolatile esters of boric acid. Before its use, the reactor was heated to 225°C. To clarify the analytical possibilities of the method, individual compounds were separated on a column (305 × 0.6 cm) containing 20% Carbowax 20 M on Chromosorb P. After eluting from the column, the carrier gas transported the compounds through the reactor with the boric acid and into the detector. The following alcohols were analyzed in this way: tert-butyl, n-butyl, benzyl, menthol, and coriandrol, as well as the following other organic compounds: p-cumene, menthylphenyl acetate, and 2-acetylpyrrol. It was shown that all of the alcohols, except coriandrol, after separation on a column, are absorbed in the reactor and are not detected; coriandrol gives a signal since it is dehydrated on the acid surface, being converted into myrcene. This method was used independently by Hefendehl [60] for the removal of terpene alcohols from terpene compounds of other classes. A reactor filled with metaboric acid (0.1 g) deposited on a support was used for the selective absorption of alcohols at 130-140°C. A layer of a mixture of calcium hydride (1 g) on a support was placed after the layer of metaboric acid in order to convert the water (which would interfere with the subsequent separation) formed in the esterification.

It was suggested [61] that for the identification and quantitative analysis of the acid components in mixtures with neutral compounds, the carrier flow should pass, after the chromatographic separation of sample components, through a reactor (100 × 0.5 cm) filled with potassium hydroxide on quartz powder (115:100). The acid components will be selectively absorbed in this reactor. By comparing the chromatograms obtained from the analytical column and from the column and the alkaline reactor, it is possible to identify and quantitatively determine the acid and neutral components of the sample. As an example, the results of the analysis of small quantities of phenol and cresols in the heavy coal tar oil were presented; it was also shown that this method is suitable for the analysis of compounds containing active hydrogen, such as indene, fluorene, pyrrole, indole, and carbazole as well as for the identification of ketosteroids and estrogens in steroid mixtures.

Complex compounds, which are sufficiently stable, are used in a number of cases to remove some of the components of the sample. Thus, for example, in the analysis of mixtures containing

BCl_3, CO, CO_2, HCl, $COCl_2$, and $SiCl_4$ by using a polar stationary
phase (dinonylphthalate) a chemical complex is formed with the
most reactive component (boron trichloride) and the remaining
components are separated on the column containing dinonylphthalate
[62]. Apparently the organic compounds of nickel, copper, and
other metals can also be used successfully for the selective absorption of amines and some other nitrogen-containing compounds, not
only for their separation [63].

The formation of hydrogen bonds between the stationary
phase and the sample components can also be used for the selective
retention of alcohols, acids, and other components capable of forming hydrogen bonds. Ackman and Burgher [64] used the property of
high-boiling alcohols to form hydrogen bonds with a polyester
phase to determine nonalcoholic impurities in alcohols.

The technique of carrying out the reactions for the formation
of nonvolatile compounds (the method of removal) is quite varied.
A simple and suitable method of functional analysis of the components of a vapor mixture was suggested by Hoff and Feit [65]. The
reactions forming nonvolatile compounds with the components of
the vapor sample took place directly in the syringe, in which a
thin film of the specific reagents was applied either on its inner
walls or on its plunger. After introducing the sample into the
chromatograph with such a syringe, the components which formed
nonvolatile compounds with the reagents were missing from the
chromatogram. Among the reagents used were metallic sodium,
sulfuric acid, aqueous bromine solution, hydroxylamine solution,
etc.

Besides group reagents, it is convenient to use specific reagents which react with only one or two components. For example,
such reagents as anhydrone, calcium chloride, phosphorus penta-
chloride, etc. (as well as molecular sieves) are used for the absorption of water, which interferes with the chromatographic analysis of many compounds. Thus, in the analysis of aqueous solutions of hydrocarbons and 3-bromo-1,1,2,2-tetrafluoropropane
[66] a reactor containing a mixture of phosphor pentoxide and 60-
80 mesh (weight ratio 9:1) was included in front of the chromatographic column (452×0.6 cm). The chromatographic separation
was carried out on a column (294×0.6 cm) containing 20% DC-710
silicone oil on firebrick. One reactor packing can be used to

analyze fifty 0.1-ml samples. The method is applicable for determining traces of 3-bromo-1,1,2,2-tetrafluoropropane.

The method of selective absorption was also used successfully
to separate isomeric compounds. The use of a selective absorbent
made it possible to analyze the isomeric bromoparaffins 2-bromo-
butane and 1-bromo-2-methylpropane [67]. The reactor was a glass
tube (diameter 0.8 cm) filled with three consecutive portions of the
following reagents: silver nitrate on diatomite (1:1) (3 cm layer),
sulfuric acid on diatomite, and disodiumphosphate on diatomite
(1 cm layer). Each layer in the reactor was separated from the
next one by a small diatomite layer. The reactor was placed in
front of the separation column (Tween 60 was used as the stationary
phase). At room temperature 2-bromobutane is quantitatively absorbed by the silver nitrate layer while 1-bromo-2-methylpropane
hardly reacts with it at all. Dawson [68] describes the use of a reactor containing silver nitrate for the absorption of dichloroethyl-
ene and dibromoethylene in the gas chromatographic and dibromoethylene in the gas chromatographic analysis of methylethyl lead
derivatives in gasolines in connection with an electron-capture detector.

The possibility of identifying cis and trans isomers of the
1,3-dienes, based on the fact that the trans isomer reacts more
quickly with a dienophyl agent (chloromaleic anhydride) than the
cis isomer, was demonstrated by Gil-Av and Herzberg-Minzly [2].
In this work it was suggested that relatively slow chemical reactions, resulting in only partial absorption of the reacting compounds
("method of partial subtraction"), be used for the identification of
the peaks of the unknown compounds. Here, the stationary phase
(chloromaleic anhydride) which reacts selectively with dienes, was
used as the reagent.

Sometimes it is necessary to separate the steps of the formation of a new reaction product and its retention by the adsorbent.
A stably adsorbed compound can be considered as nonvolatile; of
course, the term "nonvolatile compound" is only valid under conditions (adsorbent, temperature, etc.).

Ray [69] was among the first to use the reaction for the formation of nonvolatile compounds by using the method of halogena-
tion to determine impurities in ethylene. He showed that ethylene
will be removed from the sample on a column containing 40%
liquid bromine on carbon as the result of the formation of bromo-

ethylenes which will be adsorbed by charcoal. The bromination of ethylene took place in the first column (19 × 1.1 cm) containing 12.5 g bromine on charcoal while the separation of the sample components took place in the second column (40 × 0.2 cm) filled with activated charcoal. The steps of the bromination, adsorption (removal), and the subsequent chromatographic separation of the light impurities were combined in a single apparatus. The method permitted the quantitative determination of traces of hydrogen, nitrogen, carbon monoxide, methane, and ethane in ethylene. An analogous approach was also used by Janák et al. [70, 71] for the analysis of argon mixed with oxygen and other permanent gases. At conventional temperatures argon and oxygen cannot be separated on molecular sieves. Therefore, it was suggested that oxygen be quantitatively converted into water in a hydrogen flow (used as the carrier gas) on a palladium catalyst prior to the chromatographic separation on calcium zeolite, which also retains water quantitatively. In this case the recorded peak corresponded to the true argon concentration of the sample. A small reactor filled with Alusil which had the palladium catalyst deposited on its surface (0.02 g palladium in a 0.3% catalyst) was used for the catalytic combustion. The oxidation temperature was 20-70°C. The method was used for the rapid control of the argon concentration in gas-separation plants, in ammonia synthesis, in the production of hydrogen, etc.

The method of selective formation is the opposite of the method of selective removal. In this method volatile compounds are formed in fast reactions before the column, and are then separated. This method was used for the separation of the lower carbonyl compounds from the corresponding 2,4-dinitrophenylhydrazones [71] and for the separation of amines [73].

The numerous examples cited above indicate the wide possibilities of the method of forming nonvolatile compounds in gas chromatography.

The advantage of this method is the possibility of using it continuously during chromatographic analysis. In the further development of this direction, particular attention must obviously be paid to the use of new selective reactions for oxygen-containing and other heteroatomic organic compounds, and also to the development of new variations of its application (specifically, the use of volatile reagents, the use in capillary chromatography, etc.). The method of the selective isolation of components must also be developed further.

OXIDATION

Zlatkis et al. [74] used oxidation with ninhydrine (triketohydrindene) for the analysis of α-aminoacids; in this process, light aldehydes are formed containing one less carbon atom than the initial acid. The oxidation is carried out at 140°C in a glass reactor (15 × 0.6 cm) filled with ninhydrine deposited on firebrick (30% ninhydrine). One packing was used for five analyses. In order to ensure a complete reaction the amino acid is mixed with ninhydrine: one part of an aqueous ninhydrin solution is mixed with one part of a 0.28 M amino acid solution and this mixture is introduced with a hypodermic syringe into the reactor. The samples are stored in an ice bath.

The method was used successfully for the analysis of leucine isoleucine, norleucine, valine, norvaline, α-amino-n-butyric acid, and alanine.

The aldehydes formed in the reaction were separated on a chromatographic column containing a mixture of ethylene- and propylene-carbonate as stationary phase. The separated aldehydes were converted (by hydrocracking) into methane and water; the water was absorbed on molecular sieve and the methane detected.

Chekmareva et al. recommended [75] that in the analysis of hydrogen in mixtures with carbon monoxide, carbon dioxide, nitrogen, and oxygen, using helium as the carrier gas, the mixture be separated on a column containing aqueous silicic acid (which removes carbon dioxide) followed by oxidation on copper oxide. The water, formed from the oxidation of the hydrogen, is easily separated from carbon dioxide on a column with glycerine (70°C), and is detected by a katharometer.

LITERATURE CITED

1.   A. I. M. Keulemans and S. G. Perry, Nature, 193:1078 (1962).
2.   E. Gil-Av and Y. Herzberg-Minzly, J. Chromatog., 13:1 (1964).
3.   A. A. Zhukhovitskii and N. M. Turkel'taub, Gas Chromatography, Gostekhizdat, Moscow, 1962.

4. A. I. M. Keulemans and H. H. Voge, J. Phys. Chem., 63:476
    (1959).
5. R. Rowan, Anal. Chem., 33:658 (1961).
6. I. R. Klesment, S. A. Rang, and O. G. Eizen, Neftekhimiya,
    3:864 (1963).
7. G. S. Landeberg, B. A. Kazanskii, P. A. Bazhulin, G. F.
    Bulanova, A. L. Liberman, S. A. Ukholin, E. A. Mikh-
    ailova, A. F. Plate, Kh. E. Sterin, M. M. Sushchinskii, and
    G. A. Tarasova, The Determination of the Individual Com-
    position of Gasolines by Direct Distillation with the Com-
    bined Method, Izd. Acad. Nauk SSSR, Moscow, 1959, p. 77.
8. Yu. E. Lille, in "Gas Chromatography," Tr. II Vses. Konfer.,
    Izd. "Nauka," Moscow, 1964, p. 322.
9. A. G. Pankov, N. A. Dolgova, A. F. Moskvin, M. F. Knyazeva,
    V. S. Fel'dblyum, and I. B. Romanov, in "Gas Chromatog-
    raphy," Tr. II Vses. Konfer., Izd. "Nauka," Moscow, 1964,
    p. 173.
10. C. E. Döring and H. G. Hauthal, J. Prakt. Chem., 22:59 (1963).
11. B. Smith and R. Ohlson, Acta Chim. Scand., 14:1317 (1960).
12. C. J. Thompson, H. T. Coleman, R. L. Hopkins, C. C. Ward,
    and H. T. Rall, Anal. Chem., 32:424 (1960).
13. C. T. Thompson, H. J. Coleman, R. L. Hopkins, C. C. Ward,
    and H. T. Rall, Anal. Chem., 32:1762 (1960).
14. C. T. Thompson, H. T. Coleman, C. C. Ward, and H. T. Rall,
    Anal. Chem., 34:151 (1962).
15. C. J. Thompson, H. J. Coleman, C. C. Ward, and H. T. Rall,
    Anal. Chem., 34:154 (1962).
16. M. Beroza, Nature, 196:768 (1962).
17. M. Beroza, Anal. Chem., 34:1801 (1962).
18. M. Beroza and R. Sarmiento, Anal. Chem., 36:1744 (1964).
19. J. Franc and V. Kolouskova, J. Chromatog., 17:221 (1965).
20. V. G. Berezkin, Neftekhimiya, 1:169 (1961).
21. J. Franc, J. Dvoracek, and V. Kolouskova, Microchim. Acta,
    N1:4 (1965).
22. F. Drawert, R. Felgenhauer, and G. Kupfer, Angew. Chem.,
    72:385 (1960).
23. F. Drawert and K. H. Reuther, Chem. Ber., 93:3066 (1960).
24. H. G. Struppe, Chem. Techn. (Berlin), 14:114 (1962).
25. H. P. Burchfield and E. Storrs, Biochemical Applications of
    Gas Chromatography, Academic Press, New York, 1962
    [Russian translation: Izd. "Mir," Moscow, 1964].

26.   J. W. Ralls, Anal. Chem., 32:332 (1960).

27.   R. L. Stephens and A. P. Teszler, Anal. Chem., 32:1047 (1960).

28.   I. R. Hunter, J. Chromatog., 7:288 (1962).

29.   E. W. Robb and J. J. Westbrook, Anal. Chem. 35:1644 (1963).

30.   I. Hornstein, J. A. Alford, L. E. Elliott, and P. F. Crowe,
      Anal. Chem., 32:540 (1960).

31.   F. Drawert and G. Kupfer, Angew. Chem., 72:33 (1960).

32.   F. Drawert, in "Gas Chromatographie 1963," ed. H. P. An-
      gelé and H. G. Struppe, Akademie Verlag, Berlin, 1963, p. 339.

33.   M. W. Anders and G. J. Mannering, Anal. Chem., 34 : 730 (1962).

34.   M. A. Molinary, L. Lombardo, O. A. Lires, and G. J. Videlio,
      An. Asoc. Quim. Argent., 48:223 (1960); RZhKhim., 1962,
      4D206.

35.   L. V. Guild, C. A. Hollingsworth, D. H. McDaniel, and
      J. H. Watiz, Anal. Chem., 33:1156 (1961).

36   R. Dijkstra and E. A. M. F. Dahmen, Z. Anal. Chem., 181:
      399 (1961).

37.   R. F. Putnam and H. W. Myers, Anal. Chem., 34:486 (1962).

38.   J. Janák, J. Novak, and J. Sulovsky, Collect. Czechosl.
      Chem. Communs., 27:2541 (1962).

39.   M. Hager and G. Baker, Proc. Montana Acad. Sci., 22:1962
      (1963).

40.   F. E. Critchfield and E. T. Bishop, Anal. Chem., 33:1034
      (1961).

41.   J. T. Kung, J. E. Whitney, and J. C. Cavagnol, Anal. Chem.,
      33:1505 (1961).

42.   H. S. Knight and F. T. Weiss, Anal. Chem., 34:749 (1962).

43.   E. Bayer, Angew. Chem., 69:732 (1957).

44.   E. M. Arnett and P. McC. Duggleby, Anal. Chem., 35:1420
      (1963).

45.   V. G. Berezkin, A. I. Mysak, and L. S. Polak, Neftekhimiya,
      4:156 (1964).

46.   V. G. Berezkin, A. E. Mysak, and L. S. Polak, Khimiya i
      Tekhnologiya Topliv i Masel, No. 2:67 (1964).

47.   J. H. Dhont, Chem. Weekblad., 58 (N35):440 (1962).

48.   J. H. Dhont, Nature, 192:747 (1961).

49.   A. I. M. Keulemans and C. A. M. G. Cramers, in "Gas
      Chromatography 1964," ed. A. Goldup, Institute of Petrole-
      um, London, 1965, p. 154.

50.   S. Hyden, Anal. Chem., 35:113 (1963).

51. R. L. Martin, Anal. Chem., 32:336 (1960).

52. D. M. Coulson, Anal. Chem., 31:906 (1959).

53. R. L. Martin, Anal. Chem., 34:890 (1962).

54. W. B. Innes, W. E. Bambrick, and A. J. Andreatch, Anal. Chem., 35:1198 (1963).

55. V. G. Berezkin, A. I. Mysak, and L. S. Polak, in "Gas Chromatography," Tr. II Vses. Konfer., Izd. "Nauka," Moscow, 1964, p. 332.

56. A. A. Zhukhovitskii and N. M. Turkel'taub, Dokl. Akad. Nauk SSSR, 144:829 (1962).

57. D. K. Albert, Anal. Chem., 35:1918 (1963).

58. H. Kögler, Dtsch. Akad. Wiss. Kl. Chem. Geol. Biol., N1:291 (1962).

59. R. M. Ikeda, D. E. Simmons, and J. D. Grossman, Anal. Chem., 36:2188 (1964).

60. F. W. Hefendehl, Naturwiss., 51:138 (1964).

61. T. Sato, N. Shintiki, and N. Mikami, Bunseki Kagaku, 14:223 (1965).

62. S. A. Ainshtein, S. V. Syavtsillo, and N. M. Turkel'taub, in "Gas Chromatography," Tr. II. Vses. Konfer., Izd. "Nauka," Moscow, 1964, p. 270.

63. G. P. Cartoni, S. R. Lowrie, C. S. G. Phillips, and L. M. Venanzi, in "Gas Chromatography 1960," ed. R. P. W. Scott, Butterworths, London, 1960, p. 273 [Russian translation: Izd. "Mir," Moscow, 1964, p. 362].

64. R. G. Ackman and R. D. Burgher, J. Chromatog., 6:541 (1961).

65. J. E. Hoff and E. D. Feit, Anal. Chem., 36:1002 (1964).

66. E. S. Jacobs, Anal. Chem., 35:2035 (1963).

67. W. E. Harris and W. H. McFadden, Anal. Chem., 31:114 (1959).

68. H. J. Dawson, Anal. Chem., 35:542 (1963).

69. N. H. Ray, Analyst, 80:853 (1955).

70. M. Krejci and J. Janák, Collect. Czechoslov. Chem. Commun., 24:3887 (1959).

71. M. Krejci, K. Tesarik, and J. Janák, in "Gas Chromatography," ed. H. J. Noebels, R. F. Wall, and N. Brenner, Academic Press, New York, 1961, p. 255.

72. J. W. Ralls, Anal. Chem., 36:946 (1964).

73. A. T. James, A. J. P. Martin, and G. H. Smith, Biochem. J., 52:238 (1952).

74.   A. Zlatkis, J. F. Oro, and A. P. Kimball, Anal. Chem., 32:
      162 (1960).
75.   I. B. Chekmareva, V. I. Trubnikov, V. P. Pakhomov, V. G.
      Berezkin, E. S. Zhdanovich, and N. A. Preobrazhenskii,
      Vestn. Tekhn. Ékon. Inf., NIITÉKhIM, Moscow, No. 10:26
      (1964).
76.   J. Klesment, J. Chromatog., 1967 (in press).

Chapter V

Analysis of Impurities

At the present time new processes requiring pure chemical compounds are being widely introduced into industry. The detection of impurities in chemicals for and from industry has become the most important application of analytical chemistry [1]. Gas chromatography is a popular tool in these analyses, giving in a single experiment the concentrations of many impurities in a sample.

However, in applying the usual gas-chromatographic technique to the analysis of impurities in pure compounds, it is often necessary to overcome a number of difficulties: such as overloading of the column (the volume of the sample is far beyond optimum), superposition of the wide peak of the main substance on the impurity peak, deterioration of the peak separation of the impurities. Although the sample volume can be decreased by using more sensitive detectors, this does not completely solve the problem of separating the main substance from the impurities which have retention volumes close to that of the main substance, since the difference in the concentrations, and consequently also in the width of the corresponding peaks, is rather large. Therefore, one of the complex problems in the chromatographic analysis of impurities is the separation of the peaks of the main component and the peaks for impurities characterized by similar chromatographic properties. In the usual technique of gas chromatography, different devices are used [2, 3] to improve the separation: selective adsorbents, long chromatographic columns, preliminary concentration, highly sensitive detectors, etc. The determination of the impurities by means of reaction gas chromatography is possible by quantitatively

TABLE 4. Analysis of Impurities by Reaction Gas Chromatography

Reacting compounds	Change in the chromatographic characteristics of the sample components as the result of the chemical reactions		Change in the detection characteristic of the sample components as the result of the chemical reactions	
	Increase of the retention time	Decrease of the retention time	Increase in detection sensitivity	Decrease in detection sensitivity
Main substance	(a) Separation of the impurities and the main substance forming a compound of low volatility with the reagent [4, 5] (b) Frontal-chemical concentration [8, 37]			Detection of the impurities in the peak corresponding to the main substance [36]
Impurity	Concentration of the impurity by using chemical absorbents which form compounds of low volatility with the impurity [38]	Separation of the impurity and the main component by conversion of the impurities into a more volatile compound by using: (a) a tubular reactor [23]; (b) a bubbling liquid reactor [21, 22]	Conversion of the undetected impurities into compounds which are recorded by highly sensitive detectors [27, 30]	
Carrier gas	Nonselective concentration of the zones for the impurities and the main substance as the result of chemical binding of a part of the carrier gas [39]			

converting the main component and/or the impurities to obtain new compounds with chromatographic or "detection" characteristics which permit the rapid quantitative analysis of the given mixture.

In reaction gas chromatography different methods have been developed for the analysis of impurities (see Table 4).

USE OF SPECIFIC CHEMICAL REACTIONS

FOR THE RETARDATION OF THE MAIN

COMPONENT

(FORMATION OF NONVOLATILE COMPOUNDS)

In this method the main component forms a new compound with the reagent which, under the experimental conditions (temperature, reagent, etc.), is almost nonvolatile, thereby resulting in a clear separation of the impurities from the main substance. The chemical reaction occurs during any of the stages of the chromatographic analysis up to the detection.

This method was first used by Ray [4] to determine nonolefinic contaminations in ethylene. The sample being analyzed (10-25 ml) first entered a reactor (19 × 1.1 cm) filled with activated charcoal saturated with bromine (40%). The liquid products from the bromination are stably retained on the activated charcoal at room temperature. The capacity of this adsorbent for ethylene is rather large (1 g absorbs up to 60 ml of ethylene). The zone of the non-olefinic impurities (inert gases and saturated hydrocarbons) was conducted from the reactor by the carrier gas (carbon dioxide) from the reactor into the chromatographic column (40 × 0.2 cm) filled with activated charcoal. A nitrometer with an alkali solution served as the detector. The method permitted the determination of impurities in concentrations of 0.01-0.1% in ethylene. The use of a more sensitive detector would undoubtedly lower the detection limit.

In the literature, different reagents are described depending on the compound being analyzed. The main component can also be absorbed (particularly in the case when the reaction rate is small) under static conditions and not in a continuous flow before the chromatographic column. Thus, Janák and Novák [5] utilized the absorption of butadiene-1,3 (the main component) by maleic

anhydride. The sample being analyzed (50-100 ml) entered a column packed with kieselguhr and coated with maleic anhydride (30%)
with small additions of benzidine (2.5%) within 30-60 sec. The
temperature of this column was held at 100-110°C. The separation
of the unreacted impurities was carried out at room temperature
on a column containing Alusil (sodium aluminum silicate) coated
with 20% dimethylformamide. The method permitted the determination of the concentration of ethylene, propane, propylene, isobutane,
n-butane, n-butylene with isobutylene, trans butylene, and cis, butylene in butadiene-1,3. The accuracy of the analysis was 0.02% abs.

The formation of some specific bonds (for example, hydrogen
bonds [6]) can also be considered as a chemical reaction. The
merit of this type of method is the possibility of the rapid regenera-
tion of the absorbent by heating. Ackman and Burgher [7] used the
formation of such bonds to determine impurities. The higher alcohols were retained selectively by the column containing a polyester
phase while nonalcoholic impurities were rapidly eluted from the
column.

The methods described above solved only one important problem – the separation of the main substance and the impurities. However, in practice it is often necessary to analyze pure compounds
in which the impurities are present in such small concentrations
that they are not recorded by the detector.

In determining the purity of these compounds, it is necessary
to have a preliminary concentration step for the impurities. One
of the effective methods for concentrating the light impurities is
the frontal method developed by Zhukovitskii et al. [9]. A variation
of this method is the frontal chemical concentration in which, instead of an adsorbent, a chemical reagent is used. The chemical
reagent forms a new chemical compound with the main component,
which is stably retained in the reactor.

The method of frontal chemical concentration is characterized
by the following advantages: 1) there is the possibility of analyzing
all unreacted impurities, since in this case, due to the nonvolatility
of the product formed from the main component, almost all impurities become "light" and are concentrated in front of the main substance; 2) there is usually a greater capacity of the adsorbent-rea-
gent and, consequently, a larger degree of concentration for the
same column-concentrator size; 3) there is a sharp dependence of

the chemical equilibrium constant on the temperature which makes it possible, in a number of cases, to desorb the main component at relatively low temperatures.

Frontal chemical concentration of the impurities was suggested by Berezkin and Gorshunov for the analysis of hydrocarbons in carbon dioxide [37]. In this method diethanolamine was used as the reagent, which reversibly retains carbon dioxide (the main substance) while not retaining the impurities. The method makes possible the determination of hydrocarbon impurities in carbon dioxide in concentrations as low as 10 ppb.

The expansion of application area of reaction gas chromatography is associated, in particular, with the expansion of the range of chemical reactions used. Mirzayanov et al. [8] used chemical absorption for analytical purposes in determining the volatile organic impurities in hydrogen by gas chromatography. The method of frontal chemisorption concentration on palladium black, which chemisorbs hydrogen intensively at room temperature, was used in the preliminary concentration of the impurities in the hydrogen. In this method all the impurities ranging from carbon monoxide to butane can be considered as "light" with respect to the hydrogen. Under the conditions used, carbon monoxide and dioxide are hydrogenated on the palladium black to methane, permitting their detection with a highly sensitive flame ionization detector. The method makes it possible to determine organic impurities and the oxides of carbon in hydrogen in concentrations in the order of 0.1-1 ppb.

The only shortcoming of this type of method is the impossibility of determining impurities similar to the main component, since these would also react with the selective absorbent used.

CONVERSION OF THE IMPURITY
INTO A VOLATILE COMPOUND

If, as the result of the chemical reaction, a volatile compound is formed which is readily separated from the main component or from its conversion products, then the detection limit for the impurity can be sharply decreased. This method can also be considered concentration since, as the result of its application, a more concentrated zone of the impurity is usually formed.

An interesting method for the determination of the sulfur concentration in organic compounds and products was developed by Okuno, Morris, and Haines [10]. After the catalytic hydrogenation of the sample (a platinum screen is used as the catalyst) the hydrogenation products (mainly methane and hydrogen sulfide) are collected in cool traps and consequently analyzed by gas chromatography. Petroleum samples containing sulfur in the order of $10^{-2}\%$ were successfully analyzed by this method. According to the authors [10], the method can be further developed to determine sulfur concentrations in the order of parts per million.

A number of papers published on such methods are devoted to the determination of impurities in inorganic materials (in metals, peroxides, etc.).

Carpenter [11] used an analogous method for the rapid determination of carbonate impurities in inorganic substances.

The schematic of the apparatus attached to a standard chromatograph is shown in Fig. 16. Upon treating the sample being analyzed with acid (10 ml 3 N HCl), the carbonate, giving off carbon dioxide, decomposes in a closed system at a pressure of ~ 0.5 atm. During the reaction the sample is stirred vigorously for 5 min with a magnetic stirrer. After the carbonates have decomposed, the volume of the reaction flask is connected to a calibrated volume V_S which has first been evacuated and which serves as the sample loop of the chromatograph. The sample of the gas from the reaction flask contains air, water, and carbon dioxide. The water is absorbed in a drying tube with magnesium perchlorate, which is placed between the sample loop and the chromatographic column. Air and the carbon dioxide are separated at room temperature in a column (30 cm) containing silica gel at a helium velocity of 4 cm/sec. Using a katharometer detector it was possible to determine 0.01 mmoles of carbon dioxide in the sample. The method permits the analysis of samples with carbonate concentrations of 0.2 ppm. The analysis time is 10 min.

Independently of Carpenter, Jeffery and Kipping [12] also suggested a method for the determination of carbonates in ores; however, they applied a dilute orthophosphoric acid solution for the decomposition of the carbonates. The same method was used by them for the determination of carbon dioxide and nitrogen oxides in aqueous monoethanolamine solutions [13].

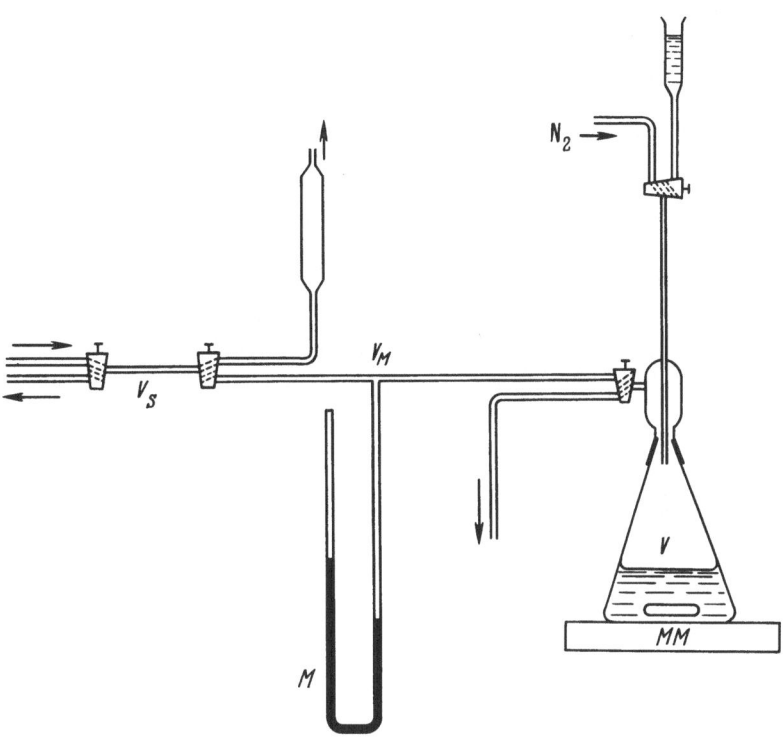

Fig. 16. Apparatus for the separation of carbon dioxide generated from carbon-ates [11]: V is the reaction volume; V_M is the intermediate volume with man-ometer M; V_S is the calibrated volume for the introduction of the gas sample into the chromatograph; MM is a magnetic stirrer.

Nelsen and Groennings [14] developed a method for the deter-mination of organic impurities in hydrogen peroxide. The deter-mination of traces of organic compounds in hydrogen peroxide is necessary in order to control its quality and stability, since an ac-cumulation of organic compounds in the peroxide can result in ex-plosive mixtures. In their method, a 10-μl sample from a glass micropipette was introduced into the upper part of a quartz tube heated to 300°C. At this temperature the sample vaporized instan-taneously and the vapors pass with helium flow into the lower part of the quartz tube which is heated to 900°C and filled with quartz rods. The decomposition of the peroxide and the oxidation of the organic substances take place simultaneously in this part of the quartz tube. The products formed – water, oxygen, and carbon dioxide – pass with the helium flow through a dryer-absorber con-

taining magnesium perchlorate, and enter the gas chromatographic
column (length, 60 cm; diameter, 6 mm) filled with silica gel. Oxygen and carbon dioxide are separated there at 30°C.

A calibration curve was used to calculate the analytical results in which the peak height of carbon dioxide was plotted against
carbon concentration. The method was used successfully to analyze
samples containing carbon at concentration levels between 1 and
200 ppm. However, the method can also be used for samples containing up to 0.5%. The relative error of the analysis was 5% (for
carbon concentrations of 100-200 ppm). The analysis time is 15
min. The method developed replaced the more-complex Liebig
method.

Juranek and Ambrova [15] developed a method for the determination of carbon, in the presence of sulfur in iron, iron alloys,
and carbides. The sample was subjected to combustion in oxygen
flow used simultaneously as the carrier gas in the chromatographic
separation of the gaseous reaction products (carbon monoxide,
dioxide, and sulfur dioxide). The separation was carried out at
room temperature on a column (5.0 × 0.5 cm) filled with silica gel.
A photocalorimetric cell [16] was used as the detector. From a
1-g sample it is possible to determine carbon concentrations in the
order of ppb. The analysis time is 25 min.

Walker and Kuo [17] described a sensitive and accurate
method for the determination of carbon in iron and its alloys. The
samples are combusted in a current of oxygen in an induction furnace. The gaseous products, after prolonged oxidation in the reactor with manganese dioxide, enter a 4-ft-long column containing a
5-Å molecular sieve and operated at temperatures close to ambient.
The oxygen peak will emerge while carbon dioxide remains adsorbed. Consequently it is eluted by programming the column temperature up to 275°C. The peak area values were measured with a disc
integrator. The method can be used for the analysis of samples in
which the carbon content varies over very broad limits, from 5
parts per million up to 20%. The analysis time is 20 min.

A method for the simultaneous determination of carbon and
sulfur in iron was also developed by Stuckey and Walker [18]. It
was used successfully to analyze samples with a carbon concentration between 0.01 and 10% and with a sulfur concentration of 0.01-0.1%.

A highly sensitive gas chromatography method for the determination of carbon (about 1 ppm) in sodium was suggested by Mungall et al. [19]. Sukhorukhov and Ivanova [20] used a flame ionization detector to determine the carbon content of metals.

The conversion of impurities into volatile products can take place either under normal conditions, or at high temperatures, by oxidation, hydrogenation, and other reactions using selective reagents.

Mysak et al. [21, 22] developed a method for the determination of traces of water in liquid hydrocarbons and in certain other organic compounds. The essence of the method is the chromatographic determination of the amount of hydrogen evolving as the result of the reaction of water with sodium aluminum hydride in diethylene glycol dimethyl ether. The method is applicable for the determination of water in hydrocarbons (paraffins, dienes, aromatics, etc.), certain oxygenated compounds (simple ethers), and other organic compounds which do not react with the sodium aluminum hydride solution.

In this method, a detector which has a high sensitivity for hydrogen is necessary; also, one has to be able to separate the vapors of the solvent, the sample, and hydrogen. In this method a thermal conductivity detector having a rather high sensitivity for hydrogen was utilized with nitrogen or argon as the carrier gas (Dimbat-Porter-Stross sensitivity: 1700 ml·mV/mg).

The carrier gas flow passing through the reactor (Fig. 17) carries the hydrogen formed in the reaction, the solvent vapor, and the sample into the chromatographic column (a copper tube, 100 cm long with a 6.0-mm outside diameter) filled with ACK-type silica gel. Pure grade argon, containing less than 0.003% oxygen was used as the carrier gas which was further purified by passing through a trap immersed in a Dewar flask containing dry ice and acetone. However, any other sufficiently effective dryer, e. g., molecular sieves, can also be used.

The sodium aluminum hydride concentration should be approximately 10 times greater than the water content of the sample. In our experiments the sodium aluminum hydride concentration was

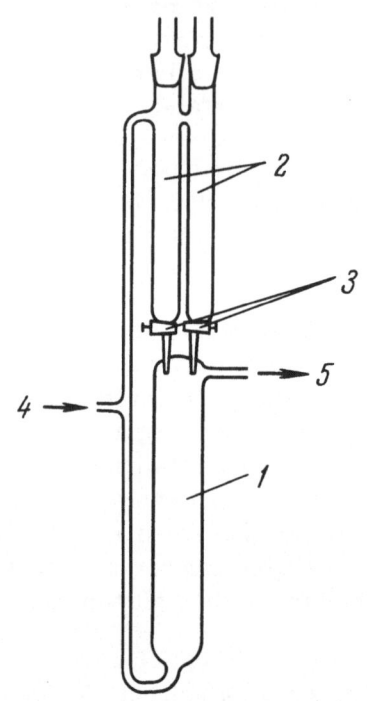

1-2%. The change in the hydride concentration within small limits does not effect the accuracy of water determination. The hydrogen peak is narrow and symmetrical; this indicates that the reaction of water dissolved in the sample with the sodium aluminum hydride solution is instantaneous and goes almost to completion.

The determination and calculation of the water concentration in the sample can be carried out by absolute calibration using a calibration curve, and comparison of two samples of known and unknown moisture content.

The above method was used to analyze the moisture content in cyclohexene, benzene, toluene, xylene, ethylbenzene, isooctane, isoprene, styrene, octane, vinyl-cyclohexane, vinylxylene, vinyl-toluene, diethyl ether, trioxane, etc. It permits the determination of water at concentrations at or

Fig. 17. Apparatus for the conversion of traces of water in hydrocarbons into hydrogen [19]: 1) reactor; 2) calibrated buret to hold the sample and the standard solutions; 3) stopcocks; 4, 5) inlet and outlet.

below 2 ppm. A single determination takes only a few minutes.

A general difficulty in trying to determine directly the water content in different technical solvents is that the component of the latter are characterized by retention times close to the retention time of water. Therefore, Bayer [23] suggested a method in which a reactor containing calcium carbide is placed upstream to the column. The sample passes through this reactor with the carrier gas before entering the column. As the result of the reaction of carbide with the water, acetylene is formed, which is eluted as a narrow, symmetrical peak, at room temperature, well ahead of all the other components of the complex solvent mixture. The method is suitable for the determination of water in concentrations as low as 10 ppm.

Drawert et al. [24] also used chemical reactions for the chromatographic determination of the water concentration in alcohol solutions and also to determine the ethanol concentration in blood.

For the determination of the moisture content in alcohol solutions, a reactor filled with a mixture of Sterchamol and calcium hydride (1 : 1) was used. Hydrogen formed in the reaction of water with calcium hydride was isolated on the chromatographic column from all of the alcohols and recorded as a narrow peak.

When analyzing the alcohol content of blood, ethanol is first converted into ethylene in order to improve separation and increase sensitivity. The dehydration takes place quantitatively at 200–300°C in a reactor containing phosphoric acid on Sterchamol (1:2) located in front of the chromatographic column.

CHEMICAL CONCENTRATION

OF THE IMPURITIES

These methods are based on selective retardation of the impurities in the sample by the formation of nonvolatile or only slightly volatile compounds and the subsequent analysis at elevated temperatures or in reactions with new chemical reagents.

The method of the determination of water in butane [25], based on the absorption of water by polyethylene glycol due to the formation of hydrogen bonds, can be considered as a typical example of this type of method. The sample passes through a trap (30.5 × 0.63 cm) held at 10°C which is filled with firebrick coated with 30% polyethylene glycol. Butane passes through the trap, and is retained only for a very short time by the polyethylene glycol. After the water is absorbed, the trap is heated to 90°C, connected to the gas chromatograph, the water is desorbed and conducted by the carrier gas (helium) into the chromatographic column (30% polyethyleneglycol-200 on a 2-ft firebrick column). In the column, water is separated from the other impurities present (methylmercaptan, benzene, etc.). Using 10-liter samples, the smallest determinable concentration was 200 ppb.

Berezkin and Gorshunov [37] developed a chemical method for the concentration of traces of carbon dioxide and hydrogen

sulfide based on the affinity of acid-type substances for forming
unstable compounds with organic bases, for example, ethanolamine,
at room temperature. As the temperature is increased, the reaction products decompose into the original compounds, and the concentrated impurities are separated into the individual components
on a column (450 × 0.4 cm) containing hexadecane stationary phase.
The minimum detectable concentration is in the order of parts per
million and the analysis time is 15-20 min.

In comparison with the usual adsorption method, chemical
concentration has the following advantages: 1) a high selectivity
(it is possible to concentrate either one or several impurities);
2) usually a higher capacity (and consequently a higher degree of
concentration); 3) a more pronounced temperature dependence of
the chemical equilibrium constant, which permits the desorbtion
of the concentrated impurities by a relatively small temperature
increase.

The method of chemical concentration of impurities is a grow-
ing field in reaction gas chromatography.

CHEMICAL CONVERSION OF IMPURITIES
AND MAIN COMPONENT IN ORDER TO CHANGE
THE SENSITIVITY OF DETECTION

The final result of a chromatographic analysis depends not
only on the separation in the column but also on the detection.

Thus, for example, the solution of an analytical problem can
be greatly simplified if, as the result of chemical conversions, it
is possible to obtain products which will pass the detector unre-
corded ("masking" of the main component) or, conversely, to convert the undetectable impurities into products for which the sensitivity of the detector is very high.

At the present time one of the most simple and convenient detectors of high sensitivity is the flame ionization detector. It permits the reliable detection of traces of organic compounds but at
the same time, it is practically insensitive to such important inorganic compounds as the oxides of carbon, oxygen, carbon disulfide,
carbon oxysulfide, water, etc. Methods have been suggested for
the preliminary quantitative conversion of these compounds into

methane or acetylene, which can be determined in very small amounts with a flame ionization detector. Knight and Weiss [26] used a reactor (30×0.5 cm) containing calcium carbide for the determination of traces of water. The acetylene formed was separated from the other C_2 hydrocarbons on a column containing a mixed stationary phase (13% dimethylsulfolane and 17% squalane). A flame ionization detector was used to detect the acetylene in the determination of trace moisture in hydrocarbons. In this case it is possible to determine the moisture content in the order of parts per million (sample volume about 0.5 ml). The shortcoming of the method is the heterogeneity of the reaction, which takes place relatively slowly, and is a possible source of error.

Schwenk, Hachenberg, and Förderreuther [27] developed a method for the determination of traces of carbon monoxide and dioxide in ethylene and propylene by using a flame ionization detector. All these compounds were separated on a column filled with activated charcoal [28]. After the chromatographic separation, these components entered, by the hydrogen carrier gas flow, a converter (35×0.6 cm) filled with Sterchamol having 10% nickel deposited on its surface. At 300°C the quantitative conversion of carbon monoxide and dioxide into methane takes place, the latter being recorded by the flame ionization detector. The smallest determinable concentration of the carbon oxides in propylene was 0.1 ppm (sample volume, 25 ml).

Porter and Volmer [29] studied the conversion reaction of carbon monoxide and dioxide into methane as used in gas chromatography in greater detail. When carrying out the hydrogenation reaction on a nickel catalyst in a stainless steel reactor (12.5×0.16 cm), they proved that, although the degree of conversion of the oxides is not influenced by it, an increase of the temperature from 206 to 266°C, improves significantly the symmetry of the methane peak. Under these conditions the hydrogenation takes place quantitatively. The areas of the chromatographic peaks for methane and carbon oxides, for the same sample size ($10 \ \mu l$), were found to be practically equal.

In this study it was also suggested that the hydrogenation reaction

$$CO + 3H_2 \rightarrow CH_4 + H_2O$$

be used to determine traces of hydrogen with a flame ionization detector.

Berezkin, Mysak, and Polak [30] suggested that oxygen be determined with a flame ionization detector after the double quantitative conversion of oxygen into methane. After the chromatographic separation, the oxygen fraction was conducted into the first converter, where the quantitative conversion of oxygen into carbon monoxide takes place on platinized charcoal (50% platinum) at 900°C. Consequently, the carbon monoxide formed enters a second converter in which a second quantitative conversion of carbon monoxide into methane takes place on a nickel catalyst. Hydrogen, required for the reduction of the carbon monoxide and for the operation of the flame ionization detector, was introduced into the gas flow just before the second converter. Pure argon was used as the carrier gas. The minimum detectable amount of oxygen was about 0.4 ppm. This method can be used to determine oxygen in hydrocarbons (in this case the hydrocarbons are not recorded) and, obviously, for the determination of moisture and other oxygen-containing compounds, since in the first reaction, oxygen-containing compounds are also converted into carbon monoxide.

Tesarik [31] developed a highly sensitive method for the analysis of carbon disulfide and carbon oxysulfide traces in gases. First, the sulfur compounds are separated from the other sample components (methane, carbon oxides, ethylene, acetylene, etc.) on a chromatographic column containing silica gel; then they are hydrogenated at 280-450°C to methane in a reactor filled with a nickel catalyst. This method was also extended to the determination of sulfur compounds (COS and CS_2) in monomers. Tesarik also reported on successful preliminary results for the analysis of hydrogen cyanide and dicyanogen with a flame ionization detector after the hydrogenation of these compounds into methane [31].

Gudzinowicz and Smith [32] suggested a new type of sensitive detection. Its principles are based on the fact that the compounds leaving the column react with a nonvolatile, radioactive compound forming a volatile, radioactive compound, which is then recorded by means of a Geiger counter. This detection method was used successfully to determine traces of inorganic oxidation agents (bromine, nitrosyl-chloride, fluorine, chlorine, etc.) in gases which reacted with the clathrate of the radioactive isotope of krypton

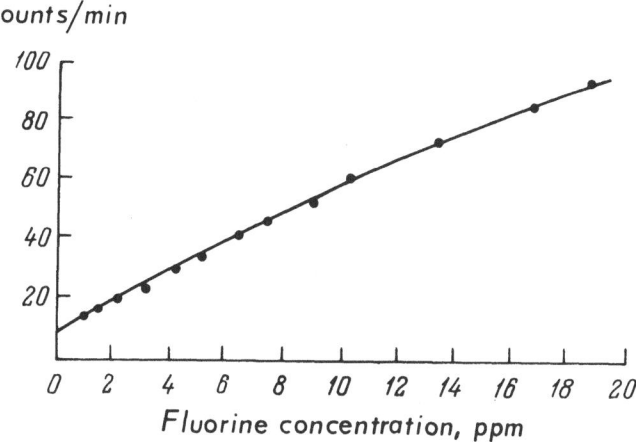

Fig. 18. Calibration curve for the radioactive detector [30].

(^{85}Kr) giving off radioactive krypton. At room temperature the clathrates of hydroquinone and ^{85}Kr can be considered stable. In order to characterize the sensitivity of this method, a calibration curve establishing the relation between the fluorine concentration and the number of counts per minute is shown in Fig. 18. As can be seen, the counting rate is directly proportional to the concentration of the oxidizing agent.

A number of papers [33-35] described detectors based on the Beilstein reaction. These detectors are used to identify the halogen-containing organic substances present in the carrier gas flow after separation in the column and general detection by a katharometer. The effluent from this detector (or a portion of it) is conducted into a burner in which a copper screen or wire is placed in the flame. The vapors of the halogen-containing compounds will color the flame green. Organic compounds containing –CN or –CCN groups give an analogous color. The detection range of this reaction is about 5-80 ppm [35].

Figure 19 shows chromatograms obtained with a flame ionization detector and a selective detector based on the Beilstein reaction. The selective detector records only the halogen derivatives; its sensitivity approximates the sensitivity of the flame ionization detector, which responds to all organic compounds. The halogen-selective detector permits the qualitative identification of the

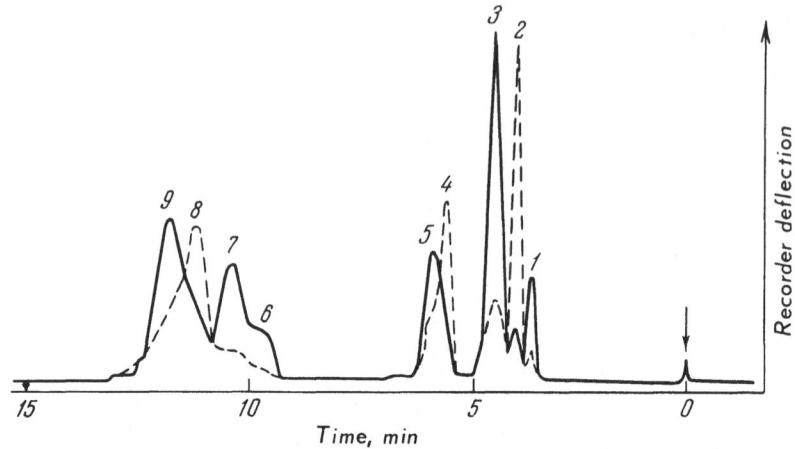

Fig. 19. Chromatograms of a complex mixture obtained by using a flame ionization detector (continuous line) and a selective detector based on the Beilstein reaction (dotted line) [33]: 1) 2,4-dimethylpentane; 2) 1,2-dichloroethane; 3) benzene; 4) 1,2-dichloropropane; 5) n-heptane; 6) 3,4-dimethylhexane; 7) 1,3-di-
methyl-cis-cyclohexane; 8) 1,2-dichlorobutene; 9) n-octane. Column: 2 m ×
5.3 mm containing 30% SF 96/1000 as the stationary phase. Carrier gas flow
rate: 66 ml/min.

impurities present in very small concentrations and the determination of the concentration of the halogenated substances even if these
compounds are not separated chromatographically from the other
organic compounds.

Selective conversion of the main component into an undetectable compound permits the detection of impurities which, in the
ordinary analysis, are masked by the peak of the main component.

Mirzayanov, Berezkin, and Nikol'skii [36] used this method
for the analysis of traces of gaseous hydrocarbons in nitrogen
oxides. The method of catalytic reduction of nitrogen and water
with palladium catalyst and in hydrogen flow was used to convert
the nitrous and nitric oxides into compounds which are not recorded by a flame ionization detector; this conversion permitted the
quantitative analysis of hydrocarbon traces which were formerly
masked by the peaks of the main components.

Gol'bert, Gorshunov, and Berezkin [39] suggested a method
for the concentration of chromatographic zones by the selective
absorption of part of the carrier gas. This method increases the

sensitivity of the subsequent detection. A mixture of helium and carbon dioxide was used as the carrier gas, the latter completely absorbed in the alkaline reactor which is placed between the chromatographic column and the detector. During the selective absorption of a part of the carrier gas the volume of the zone is decreased, thus the concentration of the substance in the zone is increased. The sensitivity of the determination increases by a factor of approximately 10.

The above-discussed examples indicate the broad prospects of the methods of reaction gas chromatography in the analysis of impurities. Reaction gas chromatography is used successfully in purity determination even in those cases when the use of conventional gas chromatographic methods does not give satisfactory results. It should also be pointed out that the application of reaction gas chromatography generally does not require a complex apparatus.

LITERATURE CITED

1. P. Auger, Contemporary Tendencies in Scientific Investigations, UNESCO, 1963.
2. A. A. Zhukhovitskii and N. M. Turkel'taub, Neftekhimiya, 3:135 (1963).
3. M. S. Vigdergaus, M. I. Afanas'ev, and K. A. Gol'bert, Uspekhi Khimii, 32:754 (1963).
4. N. H. Ray, Analyst, 80:853 (1955).
5. J. Janák and J. Novák, Collection Czechoslov. Chem. Communs., 24:384 (1959).
6. S. Dal Nogare and R. S. Juvet, Gas-Liquid Chromatography, Interscience, New York, 1962, p. 106.
7. R. G. Ackman and R. D. Burgher, J. Chromatog., 6:541 (1961).
8. V. S. Mirzayanov, V. G. Berezkin, and A. A. Datskevich, Avt. Svid. SSSR 171660 (1964); Byull. Izobr., No. 11 (1965).
9   V. S. Mirzayanov, A. A. Zhukhovitskii, V. G. Berezkin, and N. M. Turkel'taub, Zavodsk Lab., 29:1166 (1963).
10. I. Okuno, J. C. Morris, and W. F. Haines, Anal. Chem., 34:1427 (1962).
11. F. G. Carpenter, Anal. Chem., 34:66 (1962).
12. P. G. Jeffery and P. J. Kipping, Analyst, 87:379 (1962).
13. P. G. Jeffery and P. J. Kipping, Analyst, 87:594 (1962).

14.   F. M. Nelsen and S. Groennings, Anal. Chem., 35:660 (1963).
15.   J. Juranek and A. Ambrova, Collect. Czechoslov. Chem. Communs., 25:2814 (1960).
16.   J. Juranek, Collect. Czechoslov. Chem. Communs., 24:135 (1959).
17.   J. M. Walker and C. W. Kuo, Anal. Chem., 35:2017 (1963).
18.   W. K. Stuckey and J. M. Walker, Anal. Chem., 35:2015 (1963).
19.   T. G. Mungall, J. H. Mitchen, and D. E. Johnson, Anal. Chem., 36:70 (1964).
20.   O. A. Sukhorukhov and N. T. Ivanova, Zavodsk. Lab., 31(9): 1078 (1965).
21.   V. G. Berezkin, A. E. Mysak, and L. S. Polak, Khimiya i Tekhnologiya Topliv i Masel, No. 2:67 (1964).
22.   V. G. Berezkin, A. E. Mysak, and L. S. Polak, Neftekhimiya, 4:156 (1964).
23.   E. Bayer, Angew. Chem., 69:732 (1957).
24.   F. Drawert, R. Felgenhauer, and G. Kupfer, Angew. Chem., 72:385 (1960).
25.   A. A. Carlstrom, C. F. Spencer, and J. F. Johnson, Anal. Chem., 32:1056 (1960).
26.   H. S. Knight and F. T. Weiss, Anal. Chem., 34:749 (1962).
27.   U. Schwenk, H. Hachenberg, and M. Förderreuther, Brenn-stoff-Chem., 42:295 (1961).
28.   O. Horn, U. Schwenk, and H. Hachenberg, Brenstoff-Chem., 39:336 (1958).
29.   K. Porter and D. H. Volmer, Anal. Chem., 34:749 (1962).
30.   V. G. Berezkin, A. E. Mysak, and L. S. Polak, Izv. Akad. Nauk SSSR, Ser. Khim., No. 10:1871 (1964).
31.   K. Tesarik, in "Gas Chromatographie 1965," ed. H. G. Struppe and D. Obst, Akademie Verlag, Berlin, 1965, Supplement, p. 89.
32.   B. J. Gudzinowicz and W. R. Smith, Anal. Chem., 35:465 (1963).
33.   P. Chovin, J. Lebbe, and H. Moureu, J. Chromatog., 6:363 (1961).
34.   F. A. Gunther, R. C. Blinn, and D. E. Ott, Anal. Chem., 34:302 (1962).
35.   F. H. Huyten and G. W. A. Rijnders, Z. Anal. Chem., 205:244 (1965).
36.   V. S. Mirzayanov, V. G. Berezkin, and V. P. Nikol'skii, Zh. Anal. Khim., 21:1239 (1966).

37. V. G. Berezkin and O. L. Gorshunov, Izv. Akad. Nauk SSSR, Ser. Khim., No. 11:2069 (1965).
38. V. G. Berezkin and O. L. Gorshunov, Zh. Anal. Khim., 21:1487 (1966).
39. K. A. Gol'bert, O. L. Gorshunov, and V. G. Berezkin, Avt. Svid. SSSR 193777; Byull. Izobr., No. 7 (1967).

Chapter VI

Analysis of Polymers and Other Nonvolatile Compounds

Since in gas chromatography the molecules of the sample components are transported by the carrier gas flow, gas chromatography can only be used for the separation of such compounds as can be vaporized without decomposition [1, 2]. Thus, gas chromatography cannot be used directly for the separation of nonvolatile, high-molecular-weight compounds; its application is limited to volatile compounds having a vapor pressure of at least several millimeters at the temperature of the analysis.

The application of gas chromatography for the study of polymers is only possible with a basically new approach. The methods can be divided into two main groups: the analysis of polymers by the recently suggested "inverse gas chromatography," and the analysis of high-molecular-weight compounds through their breakdown products, or the products formed in chemical conversion.

In "inverse gas chromatography," in contrast to the standard method, the stationary phase is the unknown system on which known volatile compounds are separated [3]. Organic substances used as stationary phases are characterized by a given "spectrum" of relative retention times for standard compounds. This "spectrum" changes sharply when the nature of the stationary phase is changed. Polymers can also be used as unknown, stationary phases, and their molecular weight can be characterized by the changes of such "spectra." In fact, if the molecule of the polymer contains functional end groups, which differ from the main nucleus, then a fixed relation must exist between the concentration of these groups (and consequently, the molecular weight of the polymer) and the retention times of standard compounds [4].

The method of "inverse gas chromatography" can also be used to study phase transitions in polymers, e.g., in polyethylene [5]. In the region of the phase transition (fusion), a maximum can be observed in retention time and peak width plots which is due to the change in the distribution coefficient because of the phase transition.

The method of "inverse gas chromatography," obviously, can also be used to determine the structure of the polymers, the composition of copolymers, and other characteristics of the polymers which effect the change in the gas-liquid distribution constant of standard, volatile compounds. The merit of the method is the possibility of the direct investigation of the polymer (without its destruction) while using small samples; its shortcoming is that the viscosity of the compounds being studied cannot be too high at the experimental temperature [6]. The method of "inverse gas chromatography" can also be used to study the kinetics of chemical reactions [7], among them polycondensation.

At the present time an indirect method is more widely used in the study of polymers: the high-molecular-weight compound is characterized by the chromatogram of the volatile products formed during its breakdown or chemical conversion. In the study of polymers, the universal reaction of pyrolysis has the broadest use since the utilization of other chemical methods is often limited by chemical inertness, absence of functional groups, and the insolubility of most polymers having a practical importance.

In pyrolysis gas chromatography, after the pyrolysis of the nonvolatile compounds has been carried out, the volatile products formed are analyzed by gas chromatography. Usually both processes (pyrolysis and chromatographic analysis) are combined in a single apparatus.

The method of thermal decomposition, which is used in pyrolysis-gas chromatography, is one of the oldest chemical methods for studying matter. The first studies of high-molecular-weight compounds by this method were started about one hundred years ago [8]. The analytical use of pyrolysis is based on the known chemical rule that the structure and composition of a chemical compound determines its reactivity and consequently, the quantitative and qualitative composition of the products which are formed.

The shortcomings of pyrolysis are the complexity of the chemical reactions in thermal destruction and the large influence of secondary reactions. The processes of the thermal destruction of polymers have not been sufficiently studied up to the present time. Therefore, generally, it is impossible to make a prediction concerning the qualitative and quantitative composition of the volatile products formed, even if the structure of the polymer and the conditions of the pyrolysis are known; neither has the reverse problem been solved (establishing the structure and composition of the polymer from the pyrolysis products), which would be of great analytical interest.

The problem of the analyst consists of the empirical correlation of the structure of the polymer and the chromatogram of the products formed in the pyrolysis. Since the composition of the pyrolysis products is determined by many factors (temperature, duration of the pyrolysis, sample size, etc.) no generalization concerning the relationship between the pyrolysis products and the polymer is possible until the conditions of the pyrolysis, and also some standard samples, are standardized.

Thermal destruction and the subsequent analysis of the breakdown products had been used for the qualitative and quantitative analysis of high-molecular-weight compounds and for the determination of their structure [8-11]. The application of gas chromatography for the analysis of the pyrolysis products increase greatly its real value in polymer studies, since the thermal breakdown results in a very complex mixture of volatile products, but a few of them are characteristic of the given type of polymer. Therefore, generally, gas chromatography is better suited to the analysis of volatile compounds than spectroscopy and is simpler than mass spectroscopy.

The advantages of using gas chromatography for the analysis of the pyrolysis products are related to the following characteristics of gas chromatography:

1) the utilization of sensitive detectors makes it possible to use small samples of the polymers (in the milligram or even microgram range) for the analysis;

2) the high efficiency of the packed (1–3000 theoretical plates) or capillary (20–60,000 theoretical plates) columns;

3) the analysis time is short, usually varying between a few tenths and several minutes;

4) a standard chromatographic apparatus is used for the separation and the pyrolysis unit is made in the form of a small, independent attachment to the standard chromatograph.

The method of studying nonvolatile compounds from the chromatogram of the volatile breakdown products is often called pyrolysis gas chromatography.

In the first reports on the use of gas chromatography for the analysis of the thermal breakdown products of polymers [12–14], the pyrolysis was carried out in a special apparatus and the products were collected and analyzed in a second step in a standard gas chromatograph. This method is convenient in the case when small samples (of the order of 1–10 mg) cannot be used due to the heterogeneity of the samples or when the mechanism and kinetics of the pyrolysis, estimate of the thermal stability of the material, or the composition of the products at low degrees of conversion, etc., are studied [15].

In the most widespread type of pyrolysis gas chromatography, the processes of pyrolysis and chromatographic separation are organically united in one apparatus: the sample is pyrolyzed in a special reactor placed in front of the gas chromatographic column, in the stream of the carrier gas. This method was first described by Lehrle and Robb [16]. It has a number of advantages, namely: the time of analysis is decreased since pyrolysis and the introduction of the sample into the gas chromatograph are combined into a single, short-term operation; only small amounts of the sample (of the order of several milligrams or even micrograms) are needed for the study; in case of small samples (thin layers), the possibility of secondary reactions is decreased; the pyrolysis can be accomplished readily under reproducible, controlled conditions, and it is easy to analyze gaseous as well as liquid products.

An interesting, intermediate variation of the pyrolytic method was suggested by Swann and Dux [17]. The pyrolysis of the polymer was carried out in an evacuated glass ampoule (sample weight, 50 mg; heating time, 15 min). The pyrolysis products were analyzed by gas chromatography after breaking the ampoule in the carrier gas flow, at the inlet of the column. An analogous method was used successfully for the quantitative analysis of hydroxyethyl groups in hydroxyethyl starch [18]. The sample (0.001 g) was pyrolyzed in a closed capillary (9.0 × 0.1 cm) in vacuum at 400°C for 10 min. The pyrolysis products were separated by gas chromatography. A linear relationship was established between the height of the acetaldehyde peak and the quantity of hydroxyethyl groups in the sample analyzed.

In the most widely used pyrolytic methods the pyrolysis chamber simultaneously serves as the sampling system for the chromatographic column. Thus, the apparatus for characterizing nonvolatile compounds by the chromatogram of their pyrolysis products consists of the chromatograph and the pyrolysis chamber (reactor) placed in front of the chromatographic column.

The instrumentation of pyrolysis gas chromatography is distinguished by a great diversity. The construction of the pyrolysis chambers differs one from the other; they can, however, be divided into two types. In the first type, the pyrolysis takes place on a filament rapidly heated with an electrical current. The heated spiral is placed in a flow chamber the walls of which have a temperature which does not exceed the temperature of the consecutive chromatographic separation.

The substance being studied is deposited on the metallic (usually platinum or Nichrome) filament. The spiral coated with the substance is introduced into the chamber which is hermetically closed, and the carrier gas flow is started. After the gas chromatograph is set to the required analytical conditions, the spiral is rapidly heated electrically. The pyrolysis products are carried off by the carrier gas into the chromatographic column, separated and recorded by the detector. Such pyrolysis chambers are distinguished by their simple construction, the possibility for rapid heating of the sample to a given temperature, the fact that the pyrolysis is carried out in the stream of the carrier gas, and the low surface heating which decreases the secondary reactions of the pyrolysis products. Chambers of this type have the following requirements concerning their construction: 1) the volume of the chamber should be as small as possible, since an increase in the volume results in a decrease in the efficiency of the consecutive chromatographic

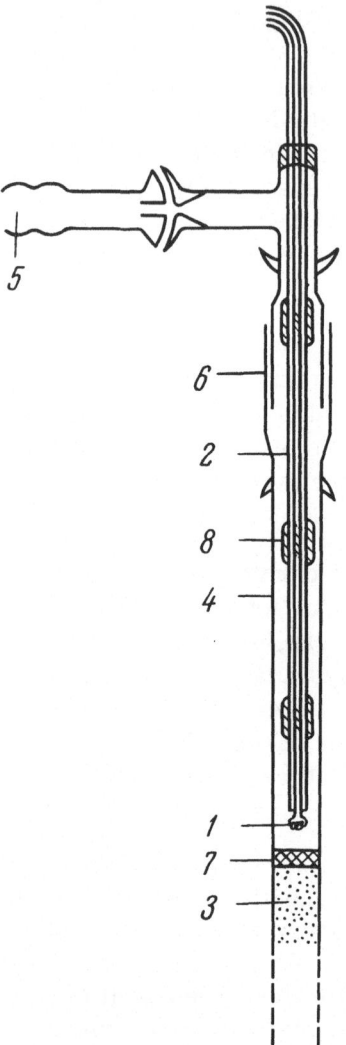

separation; 2) the chamber's walls should have an additional heating in order to avoid the possible condensation of liquid pyrolysis products on the cool walls of the chamber (another possibility is to place the whole chamber in a thermostat); 3) it should be possible to rapidly and conveniently replace the spiral coated with the sample.

Pyrolytic chambers of this type can be used for carrying out pyrolysis in different, among them corrosive, gas media, or even in a vacuum.

A typical construction of a glass pyrolysis chamber [19] is shown in Fig. 20. Cells with a similar type of construction were described among others by Janák [20], Jones and Moyles [21, 22], and Mljenek [23].

In these reports the investigation was carried out in a chromatograph equipped with an ionization detector, since the total volume of the pyrolysis products in the case of very small samples is not sufficient for detection with the more simple but less sensitive thermal conductivity detector.

Fig. 20. Glass pyrolysis chamber [19]: 1) Nichrome spiral (0.2 ohm); 2) tungsten electrodes (diameter 1 mm); 3) column packing; 4) column; 5) carrier gas; 6) ground glass joint; 7) asbestos layer; 8) glass insulator.

Franc and Blaha [24] described a pyrolytic chamber in which a platinum grid was used instead of the spiral. This modification permitted the use of a katharometer without increasing the specific load of the polymer being analyzed, i.e., the weight of polymer per unit of area of the wire.

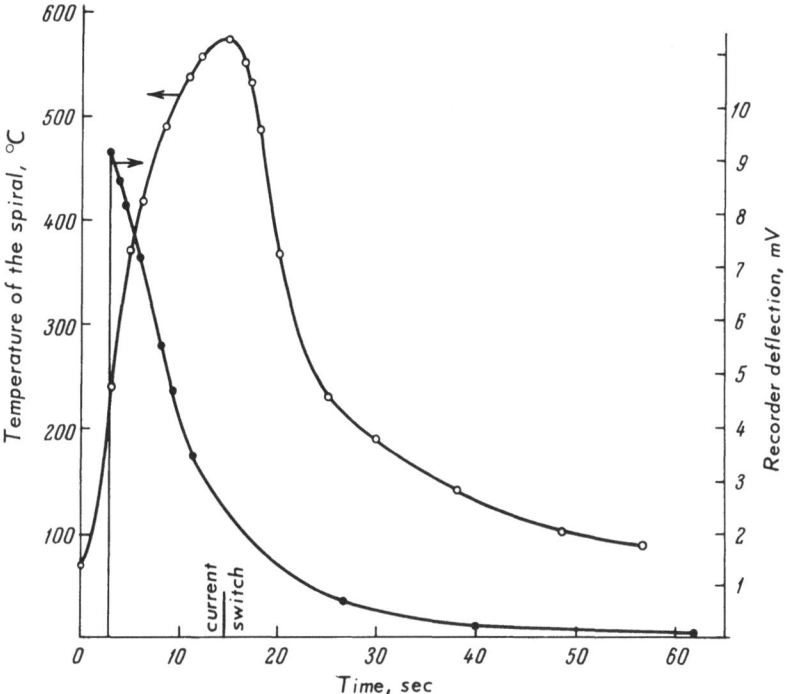

Fig. 21. The change in the temperature of the spiral and the concentration of the volatile products formed during the pyrolysis of polyropyleneoxide tetrahydrofuran.

The temperature of the cell does not remain constant in the course of the pyrolysis.

The change in the temperature of the cell spiral and the concentration of the volatile products during pyrolysis is shown in Fig. 21 on the basis of the data of Alishoeva, Kovarskii, Nemirovskii and the present author.

The evaluation of the experimental data shown in Fig. 21 shows that the pyrolysis of the polymer in such cells takes place under conditions of a rapidly changing temperature.

Kyryacos et al. [25] suggested that a higher voltage be applied to the spiral at the beginning of the pyrolysis in order to

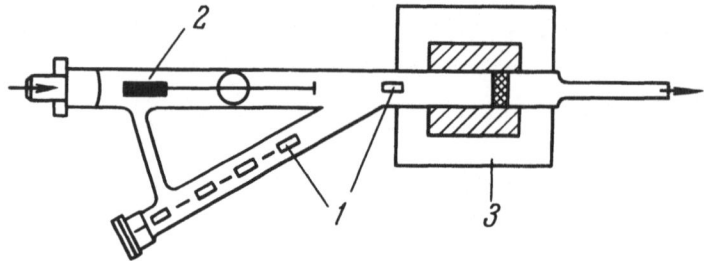

Fig. 22. Device for the pyrolysis of a polymer sample in a tubular furnace at a constant temperature [27]: 1) boat; 2) magnetic attachment
for the introduction of the sample boat into the high-temperature zone;
3) furnace.

shorten the time needed to heat the spiral to the desired temperature.

In using metallic filaments (platinum, nichrome, etc.) as the
holder for the polymer film during the pyrolysis, a catalytic activity of the metals is possible. Jones and Moyles [22] showed that
for sample amounts higher than a few milligrams a clear effect of
the filament material on the relative composition of the products
formed could be demonstrated: when using a gold-plated filament,
fewer products were formed than would be observed using a Nichrome filament. However, when smaller samples (20–30 µg) were
used, the pyrograms of polystyrene and polymethylmethacrylate
were formed to be independent of the material of the spiral (Nichrome, platinum, gold-plated platinum). Obviously this undesired
effect with large samples can be avoided if a vitrified filament is
used.

In some cases it is convenient not to place the sample directly onto the spiral but in a combustion boat of mica, quartz, or other,
more inert material.

Alishoeva et al. [26] noted that in the pyrolysis of rubber
samples in a mica boat, heated with a Nichrome spiral, more characteristic pyrograms* could be obtained (i.e., they differed more
significantly from one another) than when depositing the polymer
sample directly on the spiral.

* The term pyrogram is generally used to characterize a chromatogram obtained in
  pyrolysis gas chromatography.

In the second main type of pyrolysis instrument the sample to be pyrolyzed is introduced into a tubular reactor, heated to a high temperature, where a rapid thermal destruction takes place. The advantage of such reactors is that the thermal conditions can be better controlled and that there is less possibility that secondary reactions of the pyrolysis products will occur, because of the increased time they are present in the hot zone.

It should be pointed out that in the literature different methods are described for heating the sample to be pyrolyzed: introduction of the boat into a hot zone heated to a high temperature [27] (see Fig. 22); the use of an electrically heated platinum wire wound directly on the boat or a capillary [28, 29]; the immersion of a U-shaped chamber with the sample into a bath containing Woods alloy (500–550°C) [30]; induction heating of the sample mixed with a metal powder with a high-frequency current [31] etc.

An original system for the pyrolysis of polymers was suggested by Simon and Giacobbo [32, 33]. The sample was deposited on a wire consisting of a ferromagnetic material (0.5 × 20 mm) and placed in a glass capillary in the carrier gas stream. The ferromagnetic wire is rapidly heated to the Curie point (heating time $\sim 2 \cdot 10^{-3}$ sec) by means of a high-frequency furnace. During pyrolysis the temperature of the wire remains constant (at the Curie point). By using different ferromagnetic materials, it is possible to change the pyrolysis temperature to some degree. The method was used to study the pyrograms of different amino acids.

An increase in the size of the polymer sample being analyzed also permits, besides the gas chromatographic analyses of the volatile breakdown products, the use of other physico-chemical methods for the study of the residue (chemical and spectroscopic analysis, etc.). In practice reactors with a heated filament are used most widely.

In order to obtain reproducible results and a representative, characteristic chromatogram of the pyrolysis products, it is necessary to properly select the optimum parameters of the experiment which then can be adequately standardized.

The thermal breakdown of a polymer is sensitive to even slight changes in the pyrolysis conditions. The parameters influencing the results are the following: the size and shape of the sample; the temperature conditions and the duration of the pyrolysis; the conditions of the subsequent chromatographic separation.

Two methods are most frequently used for the deposition of the polymer on the spiral of the pyrolysis chamber: deposition from a solution by the evaporation of the solvent (for soluble substances) or from a suspension (for insoluble samples); placing small samples (several milligrams) of a solid material directly inside the spiral.

The polymer film is obtained either by submerging the spiral into a dilute solution (suspension) or the solution (suspension) of the polymer (~ 1 wt%) is deposited on one or two rings in the center of the spiral by means of a soft brush, and then dried using an infrared lamp to accelerate the drying [21].

In some cases the samples are first ground and freed of plasticizers and other dissolved substances by extraction [34]. Sometimes it is recommended that the sample be deposited on the wire from an acid or alkaline solution [33, 35].

Before the pyrolysis it is apparently convenient to heat the polymer to a slightly elevated temperature, this before the breakdown has started, in order to remove the solvent and other volatile substances (gases, monomers, moisture, etc.) dissolved in the polymers.

In studying the pyrolysis mechanism, and sometimes the detailed structure of the polymer, the application of thin films is recommended because, due to the low rate of diffusion in thick films, the role of the secondary reactions is increased. For this and also a number of other reasons, the true mechanism of the breakdown process is distorted, and the qualitative composition of the pyrolysis products is changed [21, 36].

Jones and Moyles [22] in studying the dependence of the composition of the pyrolysis products on the sample size, demonstrated the advantages of working with small (microgram) samples. In their work, two comparative chromatograms were given for a styrene homopolymer using milligram and microgram samples. The processes were carried out under identical conditions, and the appearance of additional peaks in the pyrogram when using milligram sample indicated an increased role of the secondary reactions when large samples were used. Since in practice it is diffi-

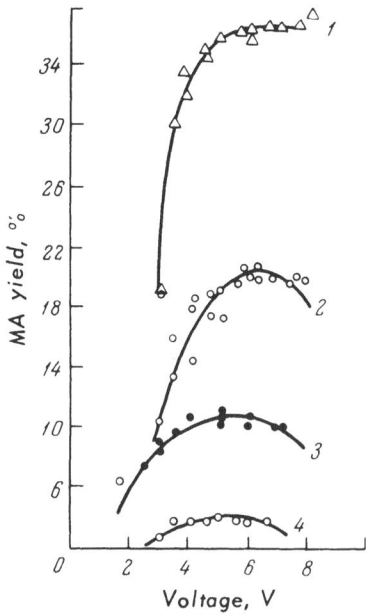

Fig. 23. The influence of the voltage of the
spiral on the yield of methyl acrylate formulation in the pyrolysis of methyl methacrylate
(MMA)—methyl acrylate (MA) copolymers (39).
1) MMA:MA = 1:3; 2) MMA:MA = 1:1; 3) MMA:
MA = 3:1; 4) MMA:MA = 9:1.

cult to obtain a thin film, the pyrolysis is usually carried out by
using thicker films which can be more easily reproduced. According to Voigt, a deviation of ±50% in the sample weight causes
no significant change in the composition of the pyrolysis products.

The possibility of using the pyrolysis method for samples
which contain other components in addition to the polymer itself is
of practical importance. In this case, by preparing aqueous emul-
sions (suspensions) of these samples they could be deposited on the
spiral and studied by the conventional method. The possibility of a
direct study of industrial samples was demonstrated in [22], where
the authors obtained identical pyrograms from the pyrolysis of the
pure polymer and of a sample containing some inert fillers. The
polymer powder was mixed with a natural resin, tragacanth (1
wt.%), and water and deposited in the form of a suspension on the

filament. The carrier gas flow rate which determines the contact time between the pyrolysis products and the spiral, which has been heated to a high temperature, can also have an effect on the results of the pyrolysis. Lehmann and Brauer noted [38] that a decrease in the carrier gas velocity from 60 to 40 ml/min doubled the benzene concentration in the products of the pyrolysis of polystyrene.

In studying the structure of the polymer, the conditions of the pyrolysis must be selected so that the pyrogram should uniquely characterize the sample. In order to fulfill this requirement it is obviously necessary that the final products be obtained in a minimum number of transformations, i.e., that they have a short "genealogy." In this case, probably, it is easier to study the relation between conversion products and the structure of the sample being studied.

In order to decrease the role of secondary reactions in the formation of the pyrolysis products, moderate temperatures (400-800°C) are usually used, and the time that the products remain in the hot reaction zone is decreased.

An increase in the number of products in the pyrolysis of polystyrene with an increase of the temperature is shown by Lehmann and Brauer [38].

The optimum pyrolysis temperature is determined by the type of polymer and the construction of the pyrolytic chamber. For example, in determining the monomer composition of a copolymer, a temperature has to be selected at which the relative yield of the characteristic pyrolysis product is not influenced by any change in the composition of the copolymer. The dependence of the yield of methylacrylate from the spiral voltage (the voltage of the spiral determines the temperature of the spiral) is given in Fig. 23 [39].

When selecting the conditions of the chromatographic separation it is necessary to take into account the nature of the pyrolysis products. Thus, for example, when studying the composition of the pyrolysis products for hydrocarbon polymers, it is convenient to use nonpolar phases, while in the study of heteroatomic compounds polar phases are recommended. In the general case, it is obviously convenient to use two or three standard chromatographic columns

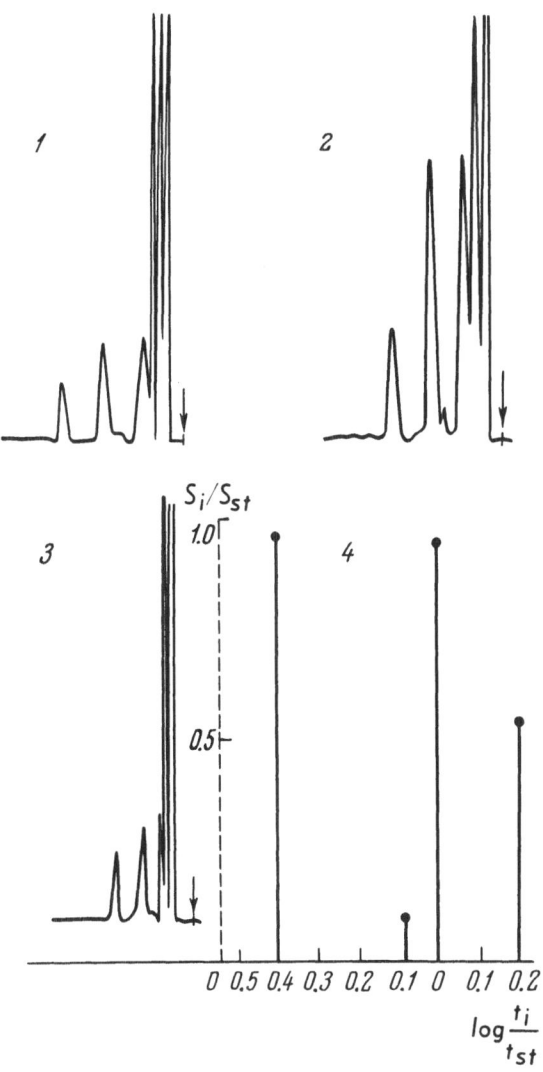

Fig. 24. The use of bar-graph-type pyrograms in pyrolysis gas chromatography: 1) pyrogram for a 2.3-mg
polymer sample and a carrier gas flow rate of 30 ml/min;
2) pyrogram for a 4.6-mg sample of the same polymer
and a carrier gas flow rate of 30 ml/min; 3) pyrogram
for a 2.3-mg sample of the same polymer and a carrier
gas flow rate of 50 ml/min; 4) bar-graph-type pyrogram
corresponding to pyrograms 1-3.

simultaneously with stationary phases which differ in polarity. Taking into considerations the broad boiling range and also the complexity of the qualitative composition of the pyrolysis products, it is desirable to utilize capillary columns with temperature programming [40], particularly when studying the mechanism of the thermal breakdown.

The results of the chromatographic analysis of the pyrolysis products can be conveniently represented in relative values by using an internal standard, and particularly by displaying these values in bar-graph-type presentation, where the ordinate and the abscissa correspond to the values

$$\frac{S_i}{S_{st}} \text{ and } \log \frac{t_i}{t_{st}} ,$$

S_i and S_{st} are the peak areas of the substance of interest and the internal standard, while t_i and t_{st} are the respective retention times. By displaying the results in this form it is possible to eliminate any possible error due to the possible nonreproducibility of sample size and shape, the carrier gas flow rate, the temperature, etc. The pyrograms obtained by Alishoeva, Kovarskii, Nemirovskii, and the present author from different sample sizes and using different carrier gas flow rates are shown in Fig. 24. All of these apparently different pyrograms correspond to a single bar-graph-type diagram.

At the present time, a large number of publications are available on discussing results obtained by investigating various polymers and nonvolatile organic substances by the method of pyrolysis gas chromatography, such as, for example, nitrocellulose, α-butylmethacrylate, polyvinly alcohol, polyvinylchloride, polymethylmethacrylate, cellulose acetate, polyethylene, polypropylene, polyacrylnitrile, natural and synthetic rubbers, barbiturates, phenothiazines, amino acids, aromatic compounds, etc.

The principle areas of the application of pyrolysis gas chromatography are discussed below.

IDENTIFICATION OF POLYMERS

As the result of numerous studies it can be concluded that when carrying out the pyrolysis under standard conditions, the

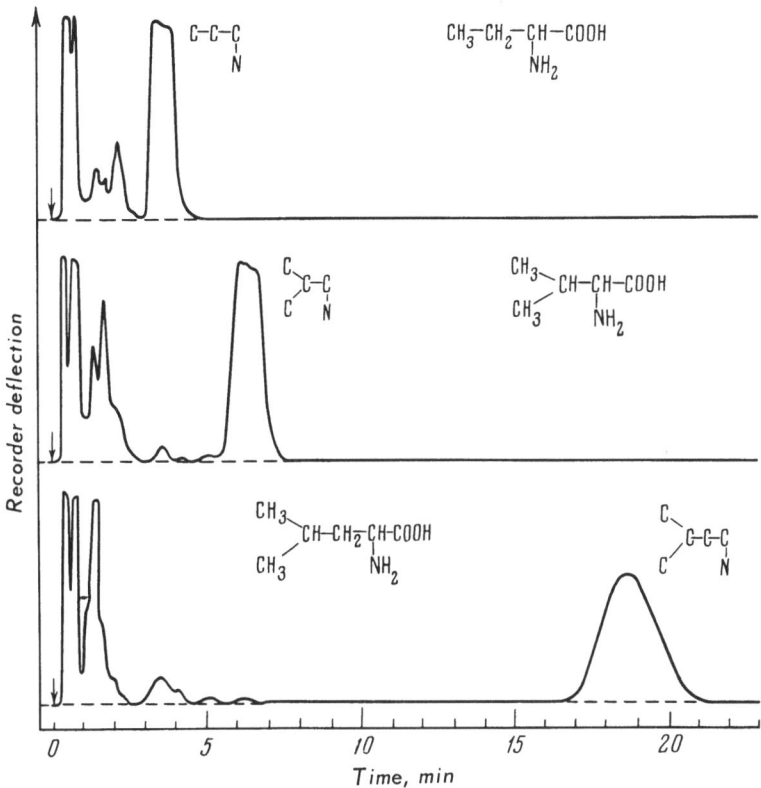

Fig. 25. Pyrograms of some amino acids [42]. Experimental conditions: stationary phase, squalane; column temperature, 30°C.

various polymers give characteristic chromatograms (pyrograms) for the pyrolysis products. Groten [29], for example, demonstrated that in testing more than 150 different polymers, almost all of the samples gave different pyrograms. Clear, characteristic pyrograms were obtained for polymers with the general formula $(CH_2 - CHX)_n$, such as: polystyrene, polyvinylacetate, polypropylene, and polyvinylchloride. The pyrograms for the cellulose esters (acetate, propionate, butyrate), for natural materials (silk, cotton, wool), and polyolefins having similar structure (polyethylene, polypropylene, poly-3-methylbutene-1, and poly 4-methyl-pentene-1) differ greatly [29].

The pyrograms obtained from a large number of plastics were studied by Nelson, Yee, and Kirk [41]. The samples (0.2-

0.5 mg) were pyrolyzed at 650-750°C for 10 sec: the pyrolysis products were separated in a column containing 5% silicone oil on Chromosorb W. The authors point out that in case of polymers with a similar structure, the only difference observed in the pyrograms was in the peaks which correspond to the "heavy" products.

Janák [20, 42] was first to use the method of identifying a substance with the help of the pyrogram of its thermal-breakdown products obtained from its pyrolysis on a heated spiral in the analysis of nonvolatile organic compounds (barbiturates, amino acids, and other biochemical materials). The chromatograms of the pyrolysis products obtained from the potassium salts of several amino acids are shown in Fig. 25. The composition of the decomposition products depends on the structure of the amino acid pyrolyzed.

Winter and Albro [43] carried out an interesting study on the pyrolytic study of proteins. The differences which were observed in the pyrograms could be related to the different structures of the substance being pyrolyzed. For example, egg and serum albumins are characterized by different pyrograms since the latter contains four times more cysteine.

Pyrolytic methods were also developed for the identification of nonvolatile aromatic compounds [24, 44], benzene clathrate [45], phosphates [46], and others. The results obtained permit us to draw the conclusion that polymers and other nonvolatile compounds can be identified from the chromatogram of their breakdown products obtained in controlled pyrolysis.

Sometimes the pyrograms obtained in pyrolysis gas chromatography are descriptively called "finger prints." This expression emphasizes the empirical nature of the method. In order to identify a polymer it is necessary to compare its pyrogram with the pyrograms of known samples and to carry out the identification of the chromatogram. Identification of the unknown sample analyzed is only possible when a pyrogram of the same material has been obtained before. Inert fillers have no effect on the polymer's pyrogram [22].

It should be noted that in some cases exact conclusions concerning the nature and structure of the polymer can only be made on the basis of the formed volatile products. Thus, for example,

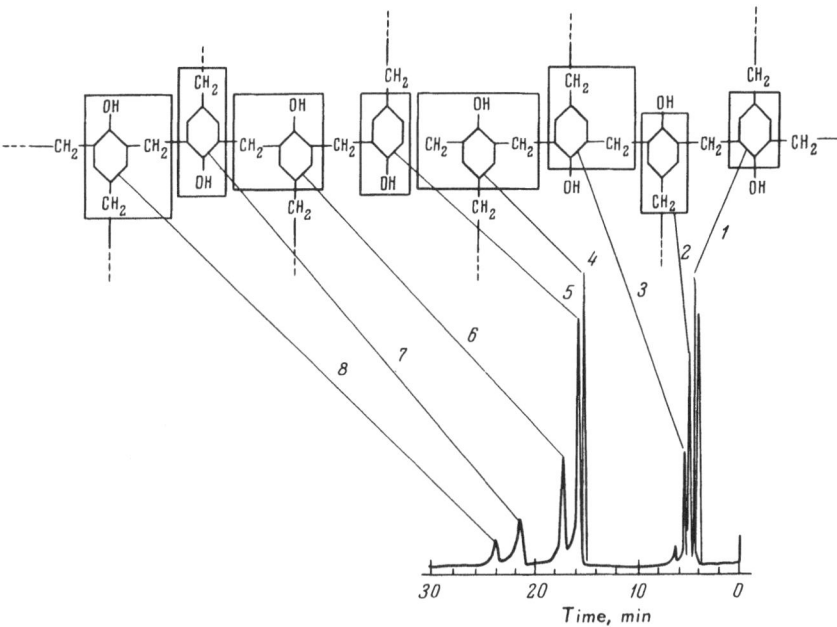

Fig. 26. Relationship between the pyrolysis products and the structure of a phenolformaldehyde resin [47]: 1) benzene; 2) toluene; 3) xylene; 4) 2,6-xylenol; 5) phenol; 6) o-cresol; 7) p-cresol; 8) 2,4-xylenol. Experimental conditions: pyrolysis temperature, 900°C; capillary column, 100 m × 1 mm i.d.; column temperature, 170°C; stationary phase, polypropyleneglycol.

by identifying the acid in the pyrolysis products of cellulose esters it is possible to determine the type of the ester which was pyrolyzed.

The individual components in the structure of condensation polymers can also be determined from the composition of the pyrolysis products, particularly if the polymer contains phenyl groups. The relationship between the chromatogram of the pyrolysis products and the structure of a phenolformaldehyde resin is well illustrated in Fig. 26 [47, 48]. The analysis of natural and synthetic fibers is described in detail in a number of publications [56–58].

The widespread use of pyrolysis gas chromatography is, in our opinion, restricted due to the absence of standard pyrolytic chambers and recommendations on standard testing conditions.

ANALYSIS OF COPOLYMERS

When studying copolymers it is essential to determine the concentration of the initial monometric units in the copolymer and the structure of the polymeric macromolecule. Within certain limitations, pyrolysis gas chromatography can give the answer to these questions.

The development of a method for the analysis of the composition of a copolymer can usually be divided into the following steps: obtaining characteristic pyrograms for copolymers of different quantitative composition; selecting characteristic peaks in the pyrogram, the relative areas of which change according to the composition of the copolymer; and finally, using these data to establish a calibration curve.

The analytical results are in satisfactory agreement with infrared spectroscopy data and permit the determination of the composition of the polymer with an error of about 1-3 abs. % [49].

The pyrolysis method was used successfully for the analysis of graft copolymers. The error in the composition in analyzing such copolymers did not exceed 5 rel. % [50, 51].

The method of pyrolysis gas chromatography is applicable to the analysis of the composition of copolymers even when the classical method of elementary analysis is limited by the close elemental composition of the initial monomers.

Esposito [52] suggested using the internal standard method in pyrolytic gas chromatography. A known amount of the polymer solution, used as the internal standard, is mixed with a known amount of the sample and diluted with a solvent. The resulting polymer solution is deposited on the filament of the pyrolysis chamber and the pyrolysis is carried out in the usual way. The volatile products are separated by gas chromatography. The concentration of the polymer (P) is determined by the formula

$$P = \frac{AB}{C},$$

where A is the area of the peak of interest, C is the area of the peak of the characteristic substance formed in the pyrolysis of the internal standard, and B is the concentration of the internal standard. Esposito used ethylmethacrylate as the internal standard;

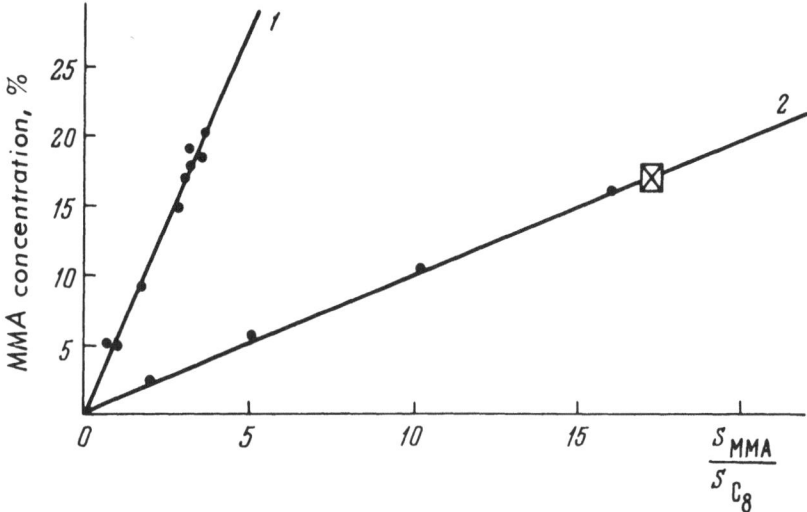

Fig. 27. The relationship between the methyl methacrylate concentration and the relative area of two characteristic peaks in the pyrograms of (1) statistical copolymers and (2) mixtures of polymethylmethacrylate and polymethylene homopolymers [53]. ⊠ corresponds to the analysis of the grafted copolymer.

it should be mentioned, however, that this method is only applicable to soluble polymers.

It must be emphasized that pyrolysis gas chromatography is not only sensitive to the composition of the copolymer, but also to its structure. This is understandable since, generally speaking, in pyrolysis the rupture of the chemical bonds does not take place only along the boundaries of the initial monomeric units. There-fore, the pyrograms of copolymers with a disordered alternation of the monomeric units (statistical copolymers) will not be identi-cal to the pyrograms of mechanical mixtures of pure homopoly-mers, and the pyrograms of graft polymers often correspond to the pyrograms of synthetic mixtures of the homopolymers of the same composition [50, 51]. This result is not unexpected since, if the number of places in the chain of the original polymer at which the grafting took place is small in comparison to the number of units of the main chain, then the pyrolytic process can be considered as the destruction of homopolymers. Two different calibration curves obtained for the statistical copolymers, methylmethacrylate and ethylene, and also for mechanical mixtures of the corresponding homopolymers, are shown in Fig. 27 [53]. In this case the point corresponding to the grafted copolymer falls on the curve of the mechanical mixture.

Pyrolysis gas chromatography is sensitive to such structural characteristics of the polymer chain as the mutual position of the substituting groups. In Groten's study [29] different pyrograms were obtained for atactic and isotactic polypropylenes – which fact underlines the great possibilities of the method.

In all of the areas indicated above, the determination of the individual composition of the pyrolysis products is fairly difficult. The problem of studying the breakdown mechanism is associated with the necessity of obtaining and separating the peaks of all of the products formed in the pyrolysis and the determination of their qualitative and quantitative composition – a very complex question. In these cases it is desirable to carry out the pyrolysis from thin layers and to use capillary columns with temperature programming for the separation of the breakdown products [40] and specific, e.g., chemical, methods for peak identification [54]. Obviously the combination of chromatography and mass spectroscopy [59] can be a great help in the identification of the chromatographic peaks.

In conclusion it should be mentioned that the literature contains sufficient experimental material to permit the recommendation of pyrolysis as a method for scientific-research studies and, in some cases, also for technological control. However, for the widespread application of this method the standardization of the apparatus and the conditions of the pyrolysis (sample size, pyrolysis temperature) and of the chromatographic separation, and further, a compilation of the pyrograms obtained in the pyrolysis of a large number of samples is necessary.

PROSPECTS FOR THE DEVELOPMENT

OF DESTRUCTION CHROMATOGRAPHIC METHODS

In studying the composition and structure of polymers, it is sometimes convenient to also apply other destruction methods, which might give more information on the structure of the samples as well as combination methods along with pyrolysis. In connection

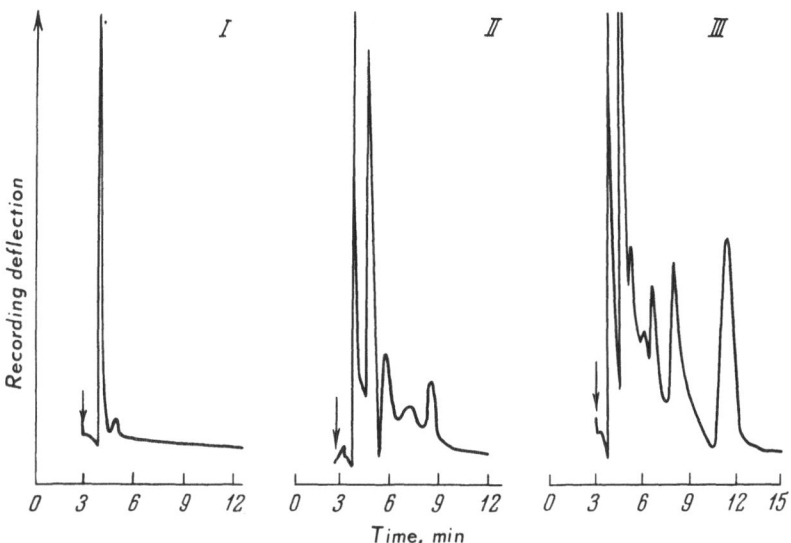

Fig. 28. Chromatograms of the volatile oxidation products of polyphenyl ethers [55]:
I) polyphenyl ether (six rings); II) convalex-10 (five rings); III) OS − 124 (five rings).

with this, pyrolysis gas chromatography should be considered as
one of the variations of destruction gas chromatography.

Scholz et al. [55] suggested the characterization of the polymers by the chromatogram of their oxidation products. The polymer to be studied is deposited on an inert, solid support and placed
in a short, stainless steel pre-column which is placed between the
sample introduction system and the analytical column. The pre-
column with the polymer can be thermostatted in the range 100–
600°C. The oxidation of the polymer takes place by introducing
1 ml oxygen into the carrier gas flow. The volatile oxidation products are carried by the carrier gas into the analytical column for
separation. The chromatogram of the volatile oxidation products
is characteristic for the polymer investigated (see Fig. 28).

Radiation chemical chromatography has been developed as a
new variation of destruction gas chromatography at the Institute of
Petrochemical Synthesis of the Soviet Academy of Science by Kol-
banov, Kyazimov, and the present author. In radiation chemical chromatography the polymer is characterized by the chromatogram of the
volatile radiolysis products. After γ-irradiation (dose, 100 Mrad)

of the polymer sample in a glass ampule which has previously been evacuated (pressure, 10^{-3} torr) the gaseous products are introduced into the gas chromatography by transferring the ampule with the irradiated polymer to the sample introduction system and breaking it there in the carrier gas flow (temperature, 50°C). About 50 samples of different polymers were studied and for each sample a characteristic chromatogram of the radiolysis products was obtained. Thus, the method can be used for the qualitative identification of polymers.

The distinguishing feature of the radiation chemical method is its high sensitivity to the hydrocarbon substituents in the main chain, which, in particular, permits the positive differentiation between high- and low-pressure polyethylenes. In the gaseous radiolysis products of high-pressure polyethylene, the ethane concentration is greatly increased (by 25 times as compared with the composition of the products for low-pressure polyethylene) and butane is also present, its yield comprising about 65% of the ethane yield.

It is interesting to follow the relationship between the chromatograms of the gaseous radiolysis products and the structure of the polymer in the series polyethylene, polypropylene polyisobutylene, which differ in the concentration and position of the methyl groups. It is convenient to characterize the radiation chemical yield of hydrocarbons by the ratio of their concentration to that of hydrogen (main product); $g_i = C_i/C_{H_2}$, where C_i and C_{H_2} are the concentrations of the hydrocarbons and hydrogen, respectively. As could be expected, the unbranched polyethylenes are characterized by a low yield of methane $gC_{H_4} = 0.3 \cdot 10^{-2} - 1.5 \cdot 10^{-2}$). The presence of a side methyl group results in a sharp increase in the methane concentration in the radiolysis products ($gC_{H_4} = 14 \cdot 10^{-2}$). In polyisobutylene two methyl groups are connected with a quaternary carbon atom, and the concentration of methyl groups in polyisobutylene is greater than in polypropylene. However, in the radiolysis products from polyisobutylene no further increase of the methane yield was observed ($gC_{H_4} = 13.5 \cdot 10^{-2}$). This result is possibly due to the low probability of the simultaneous stripping of two methyl groups from a single carbon atom.

Thus there is a relationship between the composition of the radiolysis products and the structure of the irradiated polymers.

In the case of hydrocarbon-type polymers, those light substances are produced with the highest yield which correspond to branched groups in the polymers molecule.

Besides the purely analytical application, the use of destruction methods in studying the thermal and radiation stability of polymers and nonvolatile compounds in different gaseous media is of particular interest, e.g., for the rapid determination of the effectiveness of different inhibitors. The possibility of carrying out studies with very small samples, the simplicity of the method, and its high resolving capacity uncovers broad prospects for the use of gas chromatography in the evaluation and study of polymers and other nonvolatile materials.

The development of new variations of gas chromatographic destruction methods, without a doubt, will open up new possibilities in the studies of polymers.

LITERATURE CITED

1. A. Keulemans, Gas Chromatography, Reinhold, New York, 1957 [Russian translation: IL, Moscow, 1959].

2. A. A. Zhukhovitskii and N. M. Turkel'taub, Gas Chromatography, Gostoptekhizdat, Moscow, 1962.

3. V. G. Berezkin, A. A. Zhukhovitskii, V. P. Pakhomov, L. L. Starobinets, and Z. P. Markovich, Vses. Nauchno.-Tekhn. Konf. Gaz. Khromatog., May, 1964, Moscow, 1964, p. 27.

4. V. G. Berezkin, V. P. Pakhomov, V. R. Alishoeva, L. L. Starobinets, Z. P. Markovich, and L. N. Sedov, Vysokomol. Soed., 7:185 (1965).

5. V. R. Alishoeva, V. G. Berezkin, and Yu. V. Mel'nikova, Zh. Fiz. Khim., 39:200 (1965).

6. V. G. Berezkin, V. R. Alishoeva, S. N. Ershova, and I. A. Tutorskii, Izv. Akad. Nauk SSSR, Ser. Khim., No. 9:1711 (1965).

7. V. G. Berezkin, V. S. Kruglikova, and V. E. Shiryaeva, Kinetika i Kataliz, 6:758 (1965).

8. L. A. Uoll, in "Analytical Chemistry of Polymers," Vol. II, ed. G. Klaina, Izd. "Mir," Moscow, 1965, p. 152.

9. N. Grassie, Chemistry of High Polymer Degradation Processes, Interscience, New York, 1956 [Russian translation: IL, Moscow, 1959].

10. D. L. Harms, Anal. Chem., 25:1140 (1953).

11. P. D. Zemany, Anal. Chem., 24:1709 (1952).

12. W. H. T. Davison, S. Slaney, and A. L. Wragg, Chem. &
     Ind. (London), 1356 (1954).

13. J. Haslam and A. R. Jeffs, J. Appl. Chem., 7:24 (1957).

14. G. DeAngelis, P. Ippoliti, and N. Spina, Ricerca Sci.,
     28:1444 (1958).

15. S. Glassner and A. R. Pierce, Anal. Chem., 37:524 (1965).

16. R. S. Lehrle and J. C. Robb, Nature, 183:1671 (1959).

17. W. B. Swann and J. P. Dux, Anal. Chem., 33:654 (1961).

18. H. Tai, R. M. Powers, and T. F. Protzman, Anal. Chem.,
     36:108 (1964).

19. W. H. Parris and P. D. Holland, Brit. Plastics, 33:372 (1960).

20. J. Janák, Nature, 185:684 (1960).

21. C. E. R. Jones and A. F. Moyles, Nature, 189:222 (1961).

22. C. E. R. Jones and A. F. Moyles, Nature, 191:663 (1961).

23. O. Mljenek, Chem. Prumysl., 11:604 (1961).

24. J. Franc and J. Blaha, in "Gas Chromatography 1961," Go-
     stoptekhizdat, Moscow, 1963, p. 52.

25. G. Kyryacos, H. R. Menaplace, and C. E. Boord, Anal. Chem.,
     31:222 (1959).

26. V. R. Alishoeva, V. G. Berezkin, A. A. Korolev, and I. A. Tu-
     torskii, Zh. Anal. Khim., 25:151 (1967).

27. K. Ettre and P. F. Varadi, Anal. Chem., 35:69 (1963).

28. F. A. Lehmann and G. M. Brauer, Anal. Chem., 33:673 (1961).

29. B. Groten, Anal. Chem., 36:1206 (1964).

30. E. A. Radell and H. C. Strutz, Anal. Chem., 31:1890 (1959).

31. C. C. Luce, E. F. Humphrey, L. V. Guild, H. H. Norrish,
     J. Coull, and W. W. Castor, Anal. Chem., 36:482 (1964).

32. W. Simon and H. Giacobbo, Chem. Eng. Techn., 37:1709
     (1965).

33. H. Giacobbo and W. Simon, Pharm. Acta Helv., 39:162 (1964).

34. H. Feuerberg and H. Weigel, Z. Anal. Chem., 199:121 (1964).

35. C. R. Fontan, N. C. Jain, and P. L. Kirk, Microchim. Acta,
     326 (1964).

36. J. Voigt, Kunststoffe, 51:18, 314 (1961).

37. J. Voigt, Kunststoffe, 54:2 (1963).

38. F. A. Lehmann and G. M. Brauer, Anal. Chem., 33:673 (1961).

39. R. L. Gatrell and T. J. Mao, Anal. Chem., 37:1294 (1965).

40. B. Kolb, G. Kemmner, K. H. Kaiser, E. W. Cieplinski, and
     L. S. Ettre, Z. Anal. Chem., 209:302 (1965).

41. D. F. Nelson, J. L. Yee, and P. L. Kirk, Microchem J., 6:225 (1962).

42. J. Janák, in "Gas Chromatography 1960," ed. R. P. W. Scott, Butterworths, London, 1960, p. 387 [Russian translation: Izd. "Mir," Moscow, 1964].

43. L. N. Winter and P. W. Albro, J. Gas Chromatog., 2:1 (1964).

44. J. Franc and J. Blaha, J. Chromatog., 6:396 (1961).

45. V. M. Bhatnagar and J. H. Dhont, Nature, 196:769 (1962).

46. C. E. Legate and H. D. Burnham, Anal. Chem., 32:1042 (1960).

47. Pyrolysis Accessory for Gas Chromatography (in German), Bodenseewerk Perkin-Elmer and Co., Überlingen, West Germany, 1964; cited in (48).

48. G. M. Brauer, J. Polymer Sci., No. 8(C):3 (1965).

49. J. Voigt, Kunststoffe, 54:2 (1963).

50. K. Jobst and L. Wuckel, Plaste u. Kautschuk, 12:150 (1965).

51. O. Kysel and V. Durdovič, Chem. Zvesti, 19:570 (1965).

52. G. G. Esposito, Anal. Chem., 36:2183 (1964).

53. K. J. Bambaugh, C. E. Cook, and B. H. Clampitt, Anal. Chem., 35:1834 (1963).

54. V. G. Berezkin and O. L. Gorshunov, Usp. Khim., 34:1108 (1965).

55. R. G. Scholz, J. Bednarczyk, and T. Yamauchi, Anal. Chem., 38:331 (1966).

56. O. G. Kirret and É. A. Kyullik, Izv. Akad. Nauk ÉSSR, Seriya Fiz.-Matem. Tekhn. Nauk, 13:15 (1964).

57. E. A. Kyullik and O. G. Kirret, Izv. Akad. Nauk ÉSSR, Ser. Fiz.-Matem. Tekhn. Nauk, 14:133 (1965).

58. O. G. Kirret and É. A. Kyullik, Izv. Akad. Nauk ÉSSR, Ser. Fiz.-Matem. Tekhn. Nauk, 15:252 (1966).

59. V. L. Taleroze, G. D. Tantsyrev, and V. V. Reznikov, Dokl. Akad. Nauk SSSR, 159:182 (1964).

Chapter VII

Elemental Analysis

Elemental analysis is one of the basic methods of analytical chemistry used in the characterization of compounds. The classical methods of elemental analysis ensure high accuracy; however, these methods are time-consuming and difficult to carry out. The time necessary for a single analysis of an organic compound varies between 1.5 and 3 hours [1].

The development of reliable automatic and semiautomatic devices based on classical gravimetric analytical methods is a complex task. Therefore, new accelerated physico-chemical methods of elemental microanalysis are continuously suggested and reported (see, for example, [2, 3]). One of the most promising directions in elemental analysis is the use of gas chromatography to analyze the products obtained in the conversion of the samples.

Gas chromatographic elemental analysis, in comparison with the classical gravimetric method, sometimes is less accurate; still it is quite suitable for the rapid solution of many scientific and industrial problems. In using chromatographic methods for elemental analysis the productivity is greatly increased while only relatively simple automatic or semiautomatic instruments are necessary. A number of companies are already producing instrumentation for gas chromatographic elemental analysis [3, 4].

At the present time various gas-chromatographic methods have been developed for determining the concentration of the following elements in organic compounds: carbon, hydrogen, oxygen, nitrogen, sulfur, chlorine, bromine, etc. There is no doubt as to

135

the possibility of also using gas chromatographic methods for the determination of other elements. The elemental analysis of organic compounds using gas chromatography is usually accomplished by the following steps: 1) chemical conversion of the sample into simple products: 2) chromatographic separation of these products; 3) quantitative recording of the separated products. The chemical conversions used in gas chromatographic elemental analysis represent variations of the classical methods.

The process for the conversion of the substance to be analyzed into simple "elementary" products can be carried out either in a flow-type reactor or under static conditions. The reaction in flow takes place more rapidly, and a simplified apparatus can be used. However, in this case particular attention must be paid to the completeness of the sample conversion. If the conversion process is prolonged (for example when the sample is added into the reaction zone over a prolonged period of time) or if the amount of products formed is large, the direct separation of the compounds in the chromatographic column is impossible due to the low efficiency of the separation (see, for example, [5]). In this case, prior to the chromatographic separation it is necessary to carry out the concentration of the products (e.g., by their condensation in a cool trap [6]) and to introduce them, in a separate step, rapidly into the column for separation.

In a number of cases, the primary products are subjected to a new conversion to improve separation. For example, in the analysis of the hydrogen content of the sample, acetylene, which is formed quantitatively in the reaction of water with calcium carbide, is analyzed instead of the water.

Besides the analysis of the individual compounds and technical products, methods have also been developed for both the continuous elemental analysis of compounds which have first been separated on a chromatographic column and of individual fractions obtained during the process of chromatographic separation. Such an analysis gives valuable information on the qualitative identification of the sample components.

The analytical results can be calculated either as absolute or relative values. In the first method, the weight of the compound being analyzed, the area (height) of the peaks, and the proper calibration factors (i.e., the correlation between the peak area and the concentration of the corresponding elements) have to be known. In order to increase the accuracy of the analysis, it is sometimes convenient to determine daily the calibration factors by analyzing a pure standard substance.

In the calculation of these values it is necessary to know the peak area and the proper calibration factors. This means that it is necessary to establish the relationship between the ratio of the areas of the corresponding peaks and the true ratio of the elements in the sample and standard being studied. This method is less laborious and more accurate. In some cases, in our opinion, it is possible to use a third method which incorporates the merits of both evaluation methods. In this method the material being analyzed is subjected to conversion along with an internal standard (an analogous method of using an internal standard was described in [7]).

The methods for elemental gas chromatographic analysis will be described below from different elements.

DETERMINATION OF CARBON AND HYDROGEN

In the majority of studies on the gas chromatographic determination of carbon and hydrogen, the quantitative oxidation of the organic compounds is carried out first. Oxygen, copper oxide, cobaltous-cobaltic oxide, silver permanganate, etc., are used as oxidizing agents. The separation of the products (water and carbon dioxide) is accomplished by means of different chromatographic systems and methods.

In one of the first studies [8], the oxidation of organic substances on copper oxide at 750°C in a stream of very pure oxygen was utilized. The sample (2-6 mg) in a boat was subjected to combustion in a pyrex glass tube (52 cm long; 8 mm i.d.); a tubular electrofurnace (30 cm long) was used for heating. Oxygen entered through a lateral opening at the beginning of the tube (Fig. 29). A special heater was applied around the quartz tube 6 cm from the oxygen inlet to ensure the rapid evaporation of the sample. The combustion tube was filled with the following layers of packing: platinum gauze (layer length, 6 cm); copper oxide (several 3-cm layers; total length of the copper oxide layer, 18 cm); a silverized copper screen (layer length, 6 cm). A platinum gauze (40 mesh) was used as the packing; the copper oxide was obtained as the result of the

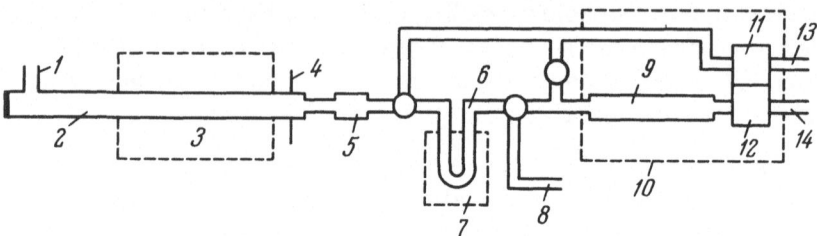

Fig. 29. Schematic of the apparatus for the elemental analysis of organic compounds [8]: 1) oxygen inlet; 2) combustion tube; 3) tubular furnace; 4) aluminum heat baffle; 5) reactor with calcium carbide; 6) freeze trap; 7) Dewar vessel with liquid nitrogen; 8) vacuum connection; 9) chromatographic column; 10) insulating jacket; 11) reference cell of the detector; 12) sensing cell of the detector; 13) helium inlet; 14) carrier gas outlet into the rotameter.

oxidation of a 40-mesh copper screen. A solution of silver nitrate was used to silverize the copper screen (size of the needles of the resulting silver crystallites, 1-2 mm). The silver layer is necessary to retard some of the oxidation products formed in the combustion of halogen- and sulfur-containing organic compounds. The incomplete removal of these products would result in the formation of additional peaks, the retention times of which are identical to the retention time for carbon dioxide [8]. The oxygen flow was first carefully dried in a tubular absorber containing magnesium perchlorate (length, 30 cm; diameter, 16 mm). The oxidation products from the organic compounds (water and carbon dioxide) were carried by the oxygen stream into the reactor (length, 5 cm) filled with calcium carbide (0.5-0.25 mm fraction), in which the water vapor is quantitatively converted into acetylene. A single reactor with a fresh calcium carbide packing was used for 2-3 analyses.

In order to concentrate the acetylene and carbon dioxide, the gas flow passed through a U-shaped glass trap, cooled with liquid nitrogen. The pressure in the trap was maintained at a level of 110 mm in order to avoid the condensation of oxygen.

After the combustion of the sample (~8 min) and concentration of the products, the trap was connected into the system of the gas chromatograph (see Fig. 29). The products that were previously condensed were introduced with a helium flow into the chromatographic column (length, 1 m; internal diameter, 6 mm) by rapidly heating the trap by removing the Dewar vessel with the liquid nitrogen. Acetylene and carbon dioxide were separated on activated silica gel (0.5-0.25 mm fraction), which was regenerated after every 80 determinations. Oxygen and nitrogen which may be present in the combustion products are easily removed on the chromatographic column from carbon dioxide and acetylene. The peak area of carbon dioxide and acetylene are proportional to the weight concentration of carbon and hydrogen, respectively, in the starting substance.

In prolonged experiments, insignificant fluctuation of sensitivity was observed on different days. Therefore the apparatus was calibrated before each series of determinations.

The method was successfully applied to such compounds as 3-aminoacetophenol, 8-hydroxyquinoline, 7-hydroxyl-7-chloraniline, benzoic acid, etc. The analysis could be carried out with an error of ±0.5 abs.% with respect to carbon and ±0.1 abs.% with respect to hydrogen. The reproducibility of the method was checked by the analysis of benzoic acid. In this case, the deviation from the mean value was ±0.36 abs. % with respect to carbon and 0.058 abs. % for hydrogen. The total analysis time, including the combustion, separation, and computation of the chromatograms is 20 min. For continuous analysis, a new sample can be introduced into the combustion unit every 10 min.

Independently of Duswalt and Brandt [8], Sundberg and Maresh [9] also developed a chromatographic method for the determination of carbon and hydrogen. In contrast to the first method [8], they used helium instead of oxygen in the combustion process, which permitted the simplification of the apparatus and, later, the determination of nitrogen. The combustion, based on the Dumas method was carried out in a current of helium using oxide as the oxidizing agent (750°C) and metallic copper for the reduction of the nitrogen oxides. In the case of substances which carbonize (e.g., sucrose), in order to ensure complete oxidation it was first mixed, before combustion, in a boat with a fine copper oxide powder. The combustion took place for at least 26 min. The water vapors were quantitatively converted into acetylene through a reaction with calcium carbide (reactor length, 20 cm; diameter, 15 cm). Carbon dioxide and acetylene were collected in a trap cooled with liquid nitrogen. The separation of the carbon dioxide and the acetylene was carried out on a column (0.9 m long; diameter, 6 mm) filled with silica gel (0.5-0.25 mm fraction).

The percent concentration of carbon and hydrogen in the samples being analyzed was calculated according to the formulas:

$$\text{for carbon}: \frac{\text{mgC}\cdot 100}{\text{sample wt., mg}} = \frac{K_{CO_2} S_{CO_2}\cdot 100}{\text{sample wt., mg}} ,$$

$$\text{for hydrogen}: \frac{\text{H}\cdot 100}{\text{sample wt., mg}} = \frac{K_{C_2H_2} S_{C_2H_2}\cdot 100}{\text{sample wt., mg}} ,$$

where S_{CO_2} and $S_{C_2H_2}$ are the peak areas, and K_{CO_2} and $K_{C_2H_2}$ are the slopes of the calibration curve for carbon dioxide and acetylene, respectively. The peak area values were determined by a planimeter.

The calibration curve for the determination of the carbon and hydrogen concentrations is given in Fig. 30 [9]. The average error was ±0.54 abs.% for hydrogen. For comparison it is pointed out that the average error of the determination of carbon and hydrogen by the gravimetric method, was ±0.33 abs.% for carbon and ±0.16 abs.% for hydrogen.

A different system was used by Vogel and Quattrone [1] for the analysis of carbon and hydrogen. They carried out the oxida-

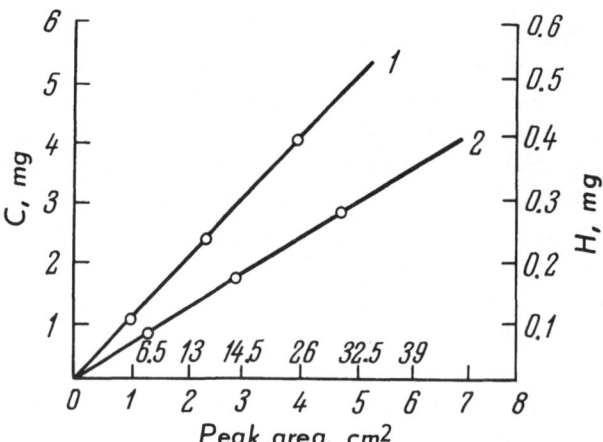

Fig. 30. Calibration curves for the quantitative determination of carbon and hydrogen [9]: 1) carbon dioxide; 2) acetylene.

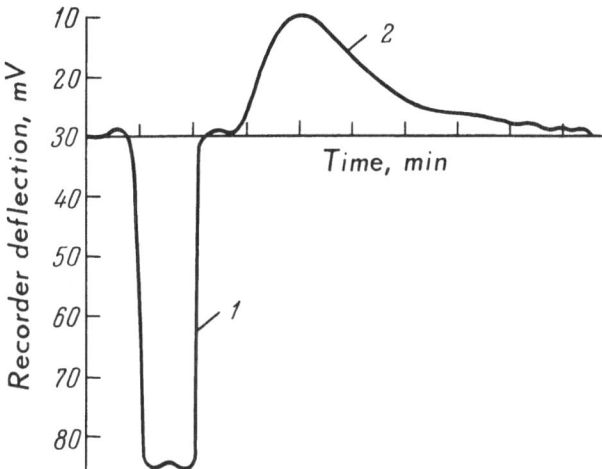

Fig. 31. Chromatogram of the conversion products in the determination of the carbon and hydrogen concentration of organic compounds [1]: 1) carbon dioxide peak; 2) water peak.

tion of the organic compounds under stationary conditions in a bomb filled with oxygen (3.8 atm pressure). The sample of the substance (8–11 mg) was placed in a boat with a platinum wire wound around which was heated by means of an electrical current (120 W). After the combustion, part of the combustion products were transferred from the bomb into a standard gas-sampling valve (25 ml) of the chromatograph. The water vapors and carbon dioxide formed were analyzed directly on the chromatographic column (2 m long) containing dodecylphthalate stationary phase, at 104°C. Dry oxygen was used as the carrier gas.

A typical chromatogram is shown in Fig. 31. Since peaks are asymmetrical, the peak areas were determined by a planimeter. The hydrogen and carbon concentrations were calculated from the peak areas.

In analyzing five different organic compounds (benzoic acid, cysteine, dextrose, glycine, β-naphthalenesulfonic acid) the average deviation was ±0.5% for carbon and ±0.8% for hydrogen. The repetitive analyses were carried out on separate days and the results were calculated on the basis of the initial calibration. A single analysis takes only 17 min. Data of three repeated determinations from a single combustion can be obtained in 40 min.

For comparison it is pointed out that the determination of carbon and hydrogen by the Pregl method takes 135-145 min (in the case of repeated analyses) and requires more skill and a more careful observance of the operational details [10].

Comparative results for the elemental analysis of samples by this method and by the method of Pregl [11] are shown in Table 5 [9]. According to the data, the mean deviation is twice as great for the gas chromatographic analysis as for the Pregl method. However, the gas chromatographic method gives greater precision in the hydrogen determination.

The method of combusting the substances in oxygen atmosphere in a bomb followed by the gas chromatographic analysis of the products was further developed by a number of researches [12-14].

The original version of the static method was suggested by Luskina et al. [15]. The necessary apparatus is relatively simple (Fig. 32). The sample was oxidized in the presence of copper oxide in a closed tube with a residual pressure of 10 mm. By using this combustion method it was possible to avoid the shortcomings of the usual oxidation method, such as contamination by uncombusted material and the possibility of explosions. In this method all the oxidation products are displaced by a helium flow and conducted directly into the chromatographic column, eliminating in this way the necessity of special sampling devices. The separation of water and carbon dioxide was carried out in helium flow on a column (0.6 m long, 3 mm in diameter), containing tricresylphosphate as the stationary phase.

TABLE 5. Comparison of the Results of Elemental Analysis Obtained by the Gas Chromatographic Method and by the Pregl Method

Element determined	Mean deviation, %		
	GC method	Pregl method	
	All samples	Benzoic acid	Ephedrine
Carbon	0.5	0.29	0.35
Hydrogen	0.84	0.33	2.1

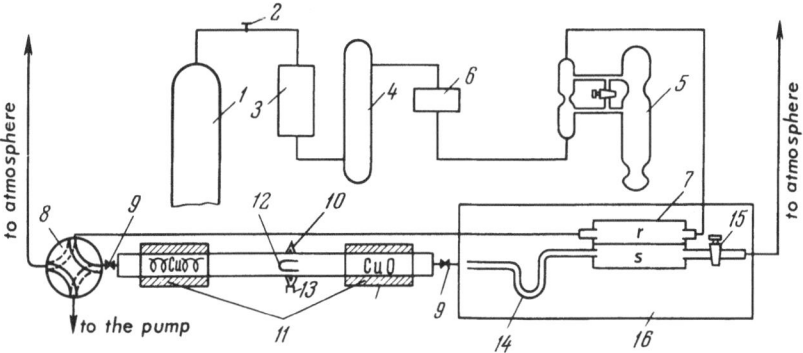

Fig. 32. Schematic of the apparatus for the elemental analysis of organic compounds [15]: 1) carrier gas source; 2) needle valve; 3) pressure regulator; 4) dryer filled with magnesium perchlorate; 5) flow meter; 6) restrictor; 7) detector (r
and s, reference and sensing cell); 8) four-way valve; 9) clamps; 10) quartz boat
with the sample; 11) electrofurnace; 12) quartz tube (35-40 cm long, 3-5 mm in
diameter); 13) gas flame; 14) chromatographic column (60 cm long, 3 mm in
diameter); 15) valve; 16) thermostat.

The carbon and hydrogen concentrations were calculated by
the usual method on the basis of calibration factors and the peak
area values. The mean deviation values were ±0.3 abs.% for carbon and ±0.15 abs.% for hydrogen.

In the publication of Vecera [16] on the automatic microdetermination of carbon and hydrogen in organic compounds, the analytical system corresponded to a standard gas chromatograph, except that instead of a standard chromatographic column chemical
absorbents were used which permitted only one product to be detected at a time. The sample (1.0-1.6 mg) mixed with cobaltous-
cobaltic oxides (30-40 mg) was rapidly burned in a quartz tube
containing the appropriate catalysts.

When determining the carbon content, the combustion was
carried out in oxygen flow. The primary combustion products
were completely oxidized by a layer of cobaltous-cobaltic oxide
deposited on corundum. After the absorption of the halides by a
layer of silver, water by magnesium perchlorate, and nitrogen
oxides by manganese dioxide, the carbon dioxide peak was detected
by a katharometer.

When determining the hydrogen concentration, the combustion
of the sample mixed with the cobaltous-cobaltic oxides was carried

out in nitrogen flow. In order to ensure the complete oxidation of
the substance, the combustion products were conducted successive-
ly, through a layer of copper oxide and cobaltous-cobaltic oxide,
and then through copper and iron layers. Iron reduces water to
hydrogen. After carbon dioxide was adsorbed in an adsorber by
anhydrone and soda asbestos, the amount of hydrogen was measured by a katharometer.

The hydrogen and carbon concentrations were calculated from
the peak areas, the sample weight, and a calibration curve.

The hydrogen and carbon dioxide peaks are shown in Fig. 33
[16]. The standard deviation is ±0.45% for carbon and ±0.16% for
hydrogen. One single analysis, including combustion and gas chromatographic analysis takes 4-6 min.

Cacace et al. [17] developed a method for the determination
of carbon and hydrogen in pure, volatile compounds already separated on a chromatographic column. The individual fractions entered with the carrier gas flow
a reactor filled with copper oxide
and iron. The conversion products
(carbon dioxide and hydrogen)
were separated at 18°C in a column (4 m long, 6 mm in diameter)
containing acetonylacetone stationary phase. Typical chromatograms are given in Fig. 34 for
benzene, diethyl ether, and cyclohexane. The method can be used
for the identification of unknown
compounds. However, its use is
limited by the relatively low accuracy of the gas chromatographic
determination of the carbon:hy-
drogen ratio. The accuracy of this
determination is about ±3% for a
single determination.

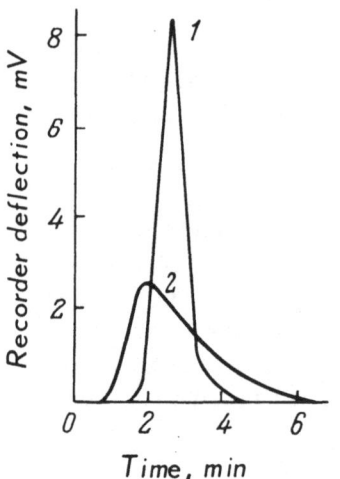

Fig. 33. Microdetermination of
carbon and hydrogen in organic
compounds [16]: 1) hydrogen peak;
2) carbon dioxide peak.

This work was extended to the analysis of labeled compounds
[18]. Conversion to hydrogen and carbon dioxide and the individual
determination of these chromatographic fractions by flow-type
proportional counters has significant advantages as compared to
the usual method for the measurement of the activity of separated

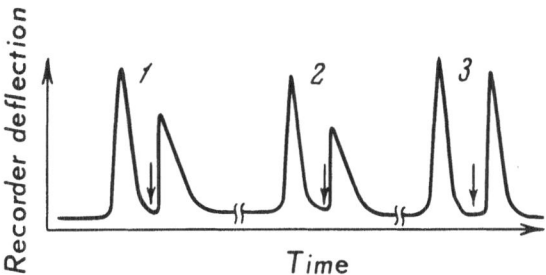

Fig. 34. Chromatograms of the elemental analysis
of previously separated components of mixtures [17]:
1) benzene; 2) diethyl ether; 3) cyclohexane.

compounds: 1) the stability of the operation of the flow-type
counter is increased since certain compounds might "poison" the
counter, impairing its efficiency; 2) the measurements are made
at room temperature, which simplifies the apparatus; 3) the simultaneous determination of the activity with respect to ^{14}C and ^{3}H is
possible for double labeled molecules.

This study is an example of the application of reaction gas
chromatography not only for elemental but also for isotropic analysis.

Skornyakov [19] suggested that compounds which have first
been separated on a chromatographic column be identified by converting them on copper oxide and iron at 700–800°C; the reaction
products (carbon dioxide and hydrogen) are separated in a different
column. The method permits the determination of the C:H ratio.

Revel'skii et al. [20, 21] also developed a negative method
for the determination of the C:H ratio and the number of carbon
and hydrogen atoms in gaseous compounds.

Apparently in some cases, particularly when analyzing impurities, it is convenient to carry out the determination of the
water formed in the combustion by coulometry [22].

A number of publications are devoted to an important special
field of elemental analysis: the determination of carbon impurities
(organic compounds) in inorganic products.

Nelsen and Groennings [23] developed a method for the determination of traces of organic compounds in hydrogen peroxide

in order to control its quality and stability. The method is based on the thermal decomposition of the peroxide and the simultaneous oxidation of the organic compounds with the consecutive chromatographic determination of the carbon dioxide formed.

The determination of the total carbon content in water by a combination of combustion and gas chromatographic methods was described by West [24].

Walker and Kuo [25] used the gas chromatographic method for the determination of carbon in iron and its alloys (range of analyzable concentrations, 5 ppw to 20%). The analysis time is 20 min.

DETERMINATION OF NITROGEN

Reitsema and Allphin [26] developed methods for the determination of the absolute concentration of nitrogen and the nitrogen: carbon ratio. The method can be used for the continuous determination of the nitrogen concentration in separated organic compounds as well as in the original samples (among them nonvolatile samples) without preliminary chromatographic separation.

Figure 35 shows the schematic of their instrument for the determination of nitrogen after chromatographic separation of the sample into components. Various values allow the individual components to be directed into the second chromatograph for the individual analysis of carbon and nitrogen. The hydrogen concentration was not determined and the water formed during combustion was

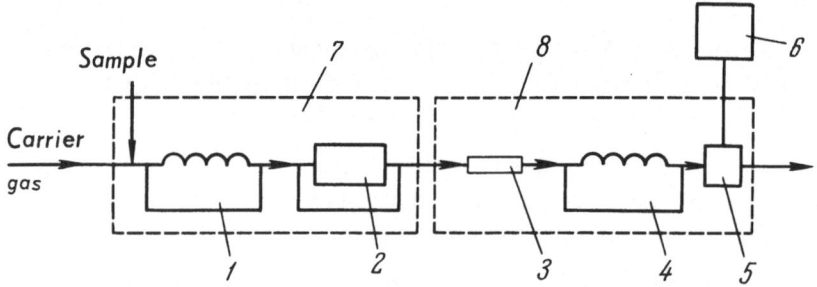

Fig. 35. Schematic of the chromatographic apparatus for the elemental analysis of nitrogen-containing compounds [26]: 1) separation column I; 2) combustion chamber; 3) dryer; 4) separation column II; 5) detector; 6) recorder; 7) thermostat I; 8) thermostat II.

TABLE 6. Instrumental Setups for the Determination of Nitrogen in Organic Substances

Use	Carrier gas	Separation column I	Separation column II
Determination of the C:N ratio in pure compounds:			
Analysis I	Helium	−	−
Analysis II	Helium	−	+
Determination of the C:N ratio of the components of a mixture:			
Analysis I	Helium	+	−
Analysis II	Helium	+	+
Determination of the absolute nitrogen concentration of a pure compound or the total nitrogen concentration in a mixture	Helium	+	−
Determination of the components of a mixture:			
Analysis I	Helium	+	−
Analysis II	Carbon dioxide	+	−

Note: The sign (+) indicates that a column is used, while (−) indicates no column used.

adsorbed in a magnesium perchlorate dryer. However, the authors pointed out the possibility of hydrogen determination by replacing the magnesium perchlorate with calcium carbide. The first thermostat is held at elevated temperatures, determined by the composition of the sample, while the second thermostat is kept at constant temperature (25-50°C). The columns and reactors are made of stainless steel. The length of the chromatographic column is 2 m, the length of the reactor for the absorption of the water is 10 cm, and the diameter of both is 6 mm. A thermal conductivity detector was used to record the results. The peak areas were measured by a printing integrator (6000 counts per minute).

The combustion of the sample was carried out by using a tubular reactor (length, 20 mm; diameter, 6 mm) filled with a dense bundle of very fine, oxidized copper wires. Prior to use, the copper wires were etched in an aqueous iron nitrate solution and then heated in air flow. The packing was regenerated every other day. Under these conditions the reactor could be used for several weeks. The optimum combustion temperature was 700°C. When

operating the reactor with fresh packing, a quantitative conversion
to carbon dioxide and nitrogen dioxide takes place; a small amount
of carbon monoxide (about 2%) was formed in cases when paraffins
were analyzed without any preliminary chromatographic separation.
The peak areas of carbon dioxide and nitrogen dioxide are proportional to the concentration of the carbon and nitrogen in the sample.
The different instrumental setups for nitrogen determination are
summarized in Table 6.

It is not necessary to weigh the sample to determine the car-
bon:nitrogen ratio. The sample is introduced into the combustion
tube, and the combustion products enter the dryer with helium flow
and are separated in the chromatographic column containing silica
gel. The ratio of the areas of carbon dioxide and nitrogen dioxide
peak is proportional to the ratio of carbon to nitrogen in the sample.
It was shown that the calibration curve depends on the operating
conditions of the separation column (Column II) such as temperature and carrier gas flow rate. The accuracy of the determination
of the carbon:nitrogen ratio is good: the maximum deviation from
the actual value is 0.15%.

This method can also be used to determine nitrogen in a
mixture which can be separated chromatographically. In such a
case, it is recommended that two analyses be carried out. In the
first analysis, the second separation column is disconnected and
the detector records the sum of the gases (carbon dioxide and nitrogen dioxide) while in the second analysis, the combustion products
are separated on the column containing silica gel. The difference
in the time of the combustion products, due to the separation in the
second column, makes it possible to compare the carbon dioxide
peak with the corresponding component of the mixture; a decrease
in the peak areas indicates the presence of nitrogen in the given
component. Moreover, the carbon:nitrogen ratio can be calculated
from the area of the carbon dioxide and nitrogen dioxide peaks
which have already been separated. However, in analyzing multicomponent mixtures, the carbon dioxide peak of one compound may
be superimposed on the nitrogen dioxide peak of the other compound. Therefore it is desirable to work out a method in which only
the nitrogen dioxide peak would be recorded. The area of this peak
would be proportional to the nitrogen concentration in the given
component and its concentration in the mixture. Such an analysis
is carried out using carbon dioxide as the carrier gas [26]. In this
case, using a katharometer, the carbon dioxide formed in the com-

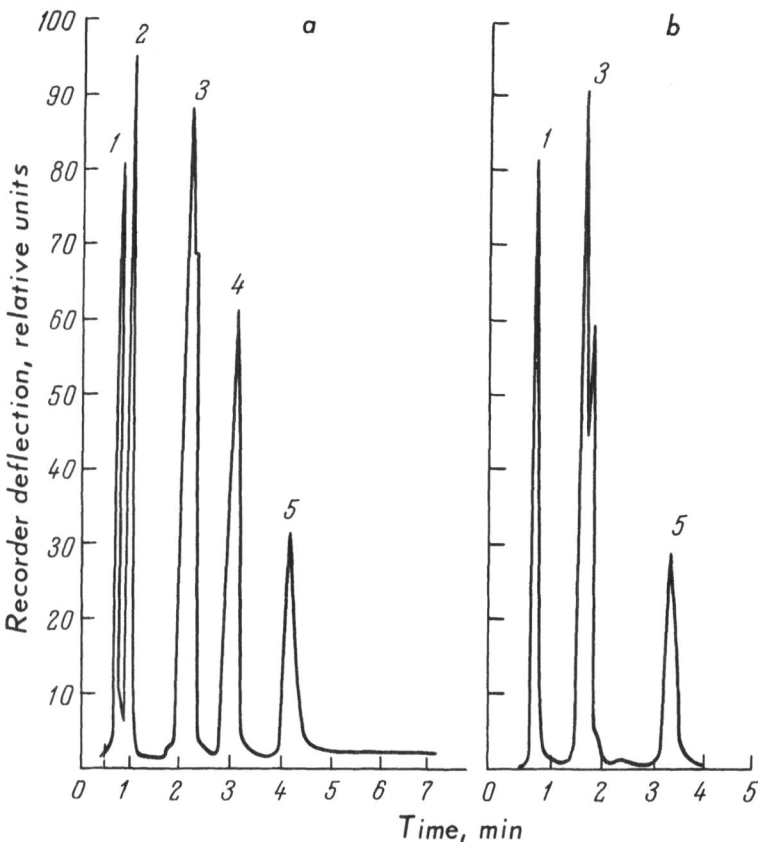

Fig. 36. Chromatograms of a mixture of hydrocarbons and nitrogen-containing compounds [26]: a) helium carrier gas; b) carbon dioxide carrier gas; 1) pyridine; 2) xylene; 3) picoline; 4) methylnaphthalene; 5) quinoline.

bustion of the organic compounds is not detected. This method can be used for the analysis of the absolute nitrogen concentration of pure compounds. In this case, the complete analysis can be carried out in 2 min. The nitrogen concentration of the analyzed compound is determined by comparing the data obtained with the results of the analysis of pure nitrogen-containing compounds carried out under analogous conditions. The volume which is analyzed is in the order of 0.001 ml. The relative error is 1–3%. It is probable that for a low nitrogen concentration in the sample ($\sim 0.03\%$) the suggested method is more accurate than the Kjeldahl method.

The indicated method was used to determine the nature of the nitrogen-containing compounds in complex mixtures. The first analysis was carried out by using helium as the carrier gas, and the second by using carbon dioxide. The chromatograms for the separation of mixtures of hydrocarbons and nitrogen-containing compounds are shown in Fig. 36 [26]. Only the nitrogen-containing compounds are recorded in the chromatogram of the second analysis.

Nightingale and Walker [27] developed a method for the simultaneous determination of carbon, hydrogen, and nitrogen with the rapid combustion of the sample by an induction furnace. The rapid heating and oxidation of the sample mixed with the catalyst in helium flow permits direct analysis of the sample oxidation products without concentration in a chromatographic column. The sample (5-25 mg), mixed with the oxidizing agent, silver permanganate (500 mg), and pieces of oxidized copper, was combusted in a quartz-lined graphite crucible by rapid induction heating for 30 sec. The oxidation products pass with a helium flow through a quartz reactor, one third of which is filled with copper oxide and two thirds with metallic copper. Such a reactor filling ensures the complete oxidation of the primary reaction products as well as the reduction of the nitrogen oxides. Consequently, the gas flow was passed through a second reactor filled with calcium carbide in which water was quantitatively converted into acetylene. The mixture of the simple compounds formed (nitrogen, carbon dioxide, and acetylene) was separated on a chromatographic column (102 cm long) containing 5-Å molecular sieve under temperature-programmed conditions. The calcium carbide converter (10.2 cm long, 6 mm in diameter) is refilled with fresh packing after each analysis cycle. All gas lines were heated. The column temperature was increased during analysis from room temperature to 400°C at a rate of 6.4°C/min. The peak areas were determined with a planimeter. The values of the calibration factor were calculated on the basis of data for the combustion of thiourea. In order to determine the retention times of the peaks corresponding to the simple products under the experimental conditions, separate determinations were carried out with standard substances. The analytical results for pure compounds of different elemental composition and structure were in good agreement with the theoretical values. The mean deviation values were: for nitrogen, $\pm 0.58\%$; for carbon, $\pm 0.52\%$; and for hydrogen, $\pm 0.22\%$. The method is also applicable to the analysis of sulfur- and halogen-containing compounds. The total analysis time is 1 hour 45 min.

Parsons, Pennington, and Walker [28] improved the method of nitrogen determination. The analysis time was shortened to 15 min.

The sample to be analyzed (1-10 mg for organic and inorganic compounds and 10-20 mg for blood) was mixed with 0.8 g of silver permanganate, a small amount of copper oxide was added, and the mixture was subjected to combustion for 30 sec using a high-frequency induction furnace.

The simple products were passed in a helium flow through a converter which was filled with a mixture consisting of 5% copper oxide and 95% reduced copper. The further oxidation of the primary products and the reduction of the nitrogen oxides to nitrogen took place in this converter. Water and carbon dioxide were absorbed with magnesium perchlorate and Ascarite; unabsorbed water was converted into acetylene in a second reactor containing calcium carbide. After elution from the column filled with the 5-Å molecular sieve, nitrogen was recorded with a katharometer. The peak area was determined with a planimeter.

The method was successfully used for the analysis of different substances, among them p-bromoacetanilide, p-nitroanilide, nicotinic acid, urea, caffeine, ammonium nitrate, sodium nitroferricyanide, 1,2,3,-benzotriazole, and blood. The accuracy of the analysis is 0.1 abs.%. In an eight-hour working day, a single worker can carry out 30 analyses.

Moskvina et al. [29] described the simultaneous determination of carbon, hydrogen, and nitrogen in organic compounds from a single sample. The conversion products, CO_2, N_2O, and N_2 were determined by gas chromatography. The combustion of the sample, mixed with copper oxide, was carried out in helium atmosphere at 650°C and a pressure of 5-10 mm. The combustion was ended in 12 min. In order to reduce the nitrogen oxides, which may form in the combustion process, a quartz tube filled with the following substances (in the order of their presence) was installed between the combustion tube and the chromatographic column: a layer of reduced copper oxide (10 cm) and a short layer of copper oxide (3-4 cm). The combustion products in helium flow were separated at 97-98°C in a chromatographic column (2 m long, 4 mm in diameter) containing triethanolamine on porous Teflon (1:1). First the chromatographic analysis of the carbon dioxide and nitrogen was carried out at a helium flow rate of 20 ml/min. Consequently, for the determination of water, the carrier gas flow rate was increased to 90 ml/min. The calculation of the nitrogen, carbon, and hydrogen concentration was carried out with help of the peak area value; the calibration factors were determined separately by the combustion of pure standard substances. Changes in the calibration factors, which were observed in individual cases, are, in the author's opinion, associated with a change in the stability of the chromatographic column. It was noted that after the packing has been replaced, the chromatographic column gave constant results only after 1-2 days of operation. The time required for a single analysis is 25-30 min. The mean square error was 0.74% for carbon, 0.22% for hydrogen, and 0.46% for nitrogen.

Pakhomova and Chumachenko [30] developed a method for the determination of nitrogen in organic compounds which was based on the oxidation of the sample with nickel oxide in a quartz test tube (16.0×1.1 cm) at 900-1100°C in helium atmosphere. The nitrogen formed was analyzed by gas chromatography. The mean deviation in the determination of the nitrogen concentration was 0.2-0.3%. The analysis took 15 min.

Gas chromatography is not only simplifying the method and shortening the analysis time, but it can also avoid some (possibly major) errors and can also be of significant help in the determination of the optimum conditions for the oxidation in the classical, gravimetric method. The possibility of such a direction in the application of gas chromatographic methods was indicated in the studies of Stewart, Porter, and Beard [31] and Hachenberg and Gutberlet [32].

As we know, when determining nitrogen by the Dumas method, too high results are obtained in many cases, particularly in the analysis of samples containing long carbon chains. Erroneous results may also be obtained if the copper oxide is depleted or if the combustion is carried out at too high a temperature.

The accurate nitrogen concentration can be obtained (even if the optimum conditions are not observed) if the whole gas collected in the absorber is analyzed by gas chromatography to establish the correction for the final results. In this case the area of the nitrogen peak will be proportional to the nitrogen concentration in the sample.

In the study made by Stewart et al. [31] the standard Coleman apparatus was used for the combustion. The sample weight was selected so that the total nitrogen volume was equal to 0.5 ml. The gas formed was completely transferred from the absorber into a gas sampling system and was analyzed at 34.5°C on a column (120 cm long) containing 5-Å molecular sieve (0.5-0.25 mm fraction). After 200 analyses the column packing had to be regenerated. The nitrogen concentration of the sample was calculated by the usual method, using the calibration factor determined in the combustion of a standard substance under identical conditions. The reproducibility of the analysis was good (the fluctuation formed was ±0.66%). The analysis time was approximately 10 min.

Hachenberg and Gutberlet [32] gave particular attention to the composition of the gas which is formed in the combustion, and determined the nitrogen concentration by introducing a correction for the concentration of other gases present, determined by the gas chromatographic method.

At first the usual analytical determination of nitrogen was carried out by the Dumas method in a semimicro scale using Unterzaucher's method [32] (the volume of the gases formed was 10-20 ml). The sample was burned in carbon dioxide flow to which oxygen has been added by bubbling the entire gas flow through a flask filled with hydrogen peroxide containing small pieces of platinum. The sample (50-100 mg) was introduced in a platinum boat into the combustion tube which contained standard filling: copper, copper oxide, and siverized cotton. The time of combustion and weighing was ~1.5-3 hours depending on the material. The time of the chromatographic analysis on two columns was 20-30 min. The following products other than nitrogen were found in the reaction products: oxygen, carbon monoxide, methane, nitric oxide, ethane, and ethylene. By correcting the final results according to the data of the gas chromatographic analysis, it was possible to greatly decrease the analytical inaccuracy. Thus, the nitrogen concentration could be determined with a mean error of ±0.02-0.05 abs.%.

The method developed had the following advantages: possibility for the accurate and rapid determination of nitrogen; increase in the accuracy and reliability of the method; possibility of carrying out the accurate analysis of compounds for which the usual analysis by the Dumas method gives poor accuracy.

Simek and Tesarik [33] used gas chromatography to determine the optimum analysis conditions. Good results were obtained
by the Dumas micromethod for the slow combustion of the sample
and by gradual increasing of the combustion temperature.

A simple and accurate method for the simultaneous determination of carbon, nitrogen, and hydrogen was developed by Walisch
[34, 35]. The sample (0.3-0.5 mg) was burned in a platinum boat
at 950°C in helium flow containing 3% oxygen. The combustion products passed through layers of copper oxide and silverized cotton.
Consequently the gas flow was directed into a reactor in which the
nitrogen oxides were reduced on a copper layer at 500°C. Here
the excess oxygen was also removed by the oxidation of the copper.
The helium flow containing carbon dioxide, nitrogen, and water
(the reaction products) passed first through a small column filled
with silica gel which adsorbed water and then, was conducted to
the first cell of the katharometer. The peak area recorded corresponded to the sum of carbon dioxide and nitrogen. Now, the gas
flow passed through a short reactor in which carbon dioxide was
adsorbed, and entered the second cell of the katharometer. The
peak area recorded was proportional to the amount of nitrogen.
Finally by rapidly heating the trap containing silica gel to 200°C,
the water was desorbed and recorded by the first cell of the katharometer. The desorption of the water was accomplished in 12
min after the introduction of the sample into the analytical system.
The relationship between the peak areas and the concentration of
the elements is linear. It was recommended that a few determinations with a standard compound be carried out every day to establish
the calibration factors. Thirty-two analyses can be made in a day.
The deviation was reported as ±0.3% for carbon, ±0.4% for nitrogen, and ±0.1% for hydrogen. It was noted that the accuracy of the
carbon determination approaches the accuracy of the classical methods, while for hydrogen the accuracy is several times better [34].

The error sources in the determination of carbon, hydrogen,
and nitrogen from a single sample using an automatic apparatus
was discussed by Clerc and Simon [36] who showed that it is possible to achieve standard deviations of ±0.13% for carbon, ±0.07%
for hydrogen, and ±0.18% for nitrogen.

DETERMINATION OF OXYGEN

The determination of the oxygen concentration in aromatic
substances often presents definite difficulties. Götz [37] developed

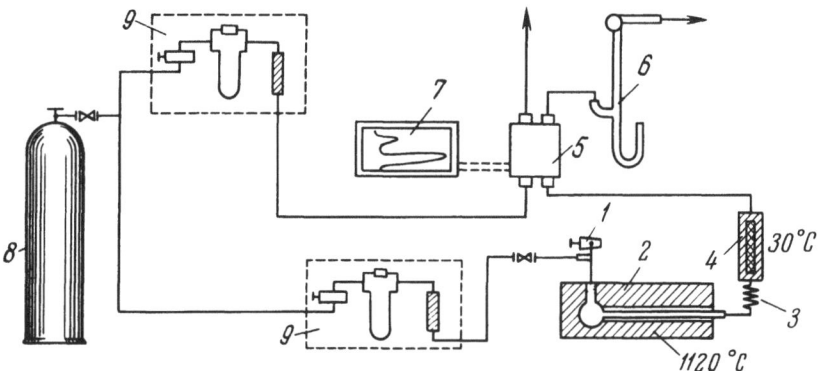

Fig. 37. Schematic of the apparatus for the determination of oxygen in organic compounds [37]: 1) sample introduction system; 2) furnace with a quartz tube; 3) metal coil for cooling; 4) chromatographic column; 5) katharometer; 6) flow meter; 7) recorder and integrator; 8) carrier gas; 9) system carrier gas regulation and purification system.

a rapid gas chromatographic method for the determination of oxygen. It is based on the cracking of the sample in a hydrogen current, the conversion of the oxygen-containing products into carbon monoxide on a graphite contact at 1120°C, and on the quantitative recording of the carbon monoxide peak after separation on a chromatographic column.

The sample was decomposed in a hydrogen flow at 1120°C (Fig. 37). The sample (2–30 mg) was introduced into the combustion chamber (1120°C) in a small platinum capsule by rotating a

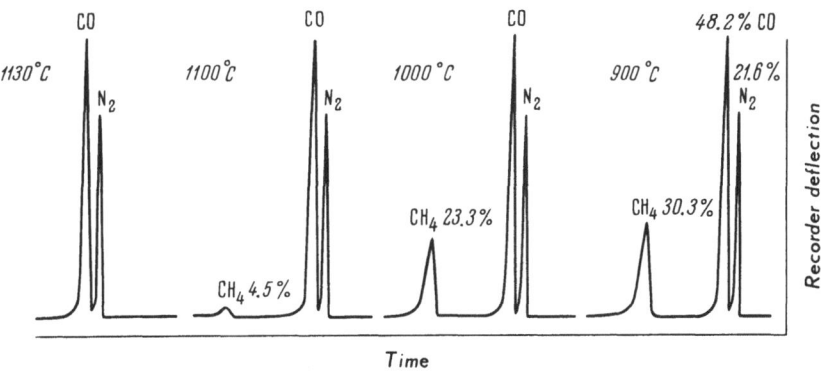

Fig. 38. Chromatograms of the products from the combustion of acetanilide with different furnace temperatures [37].

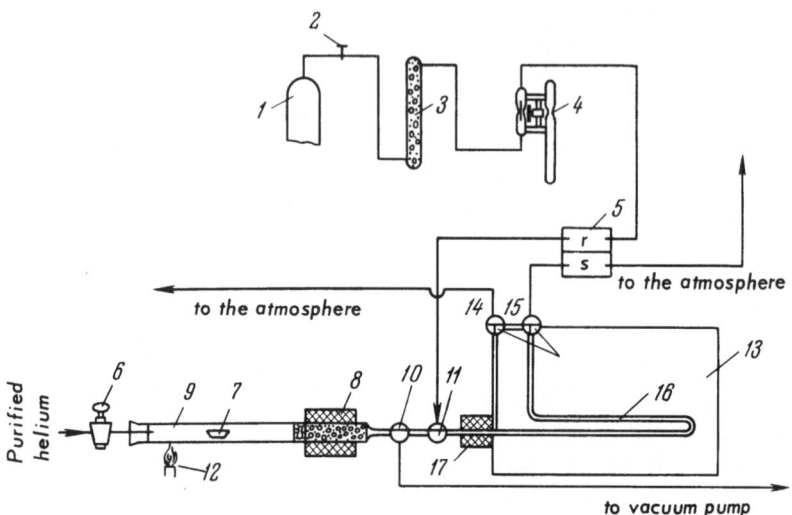

Fig. 39. Schematic of the apparatus for the determination of nitrogen, oxygen, and carbon [39]: 1) helium tank; 2) fine control valve; 3) drying tube containing ammonium perchlorate; 4) flow regulator; 5) detector (r and s are reference and sensing cells); 6) one-way stopcock for the introduction of oxygen-free helium; 7) sample boat; 8) electric furnace at 90°C; 9) quartz tube (6-7 mm in diameter; volume, 10 ml) containing "nickelized" carbon black; 10, 11) three-way stopcock; 12) flame; 13) thermostat; 14, 15) three-way stopcocks; 16) chromatographic column (60 cm long, 4 mm in diameter); 17) electric furnace (300°C).

glass stopcock. The decomposition products, in the hydrogen flow, passed in turn through a layer of quartz packing, a capillary, a layer of carbon black to SK-3 (length of the layer, 35.0 cm) and, after cooling, entered the chromatographic column (5 m long, 6 mm in diameter) filled with activated charcoal. Figure 38 shows a typical chromatogram of the products formed when analyzing acetanilide. Due to the use of hydrogen as the carrier gas, only carbon monoxide is recorded by the katharometer. The time of a single analysis is 6 min. The mean absolute error is ±0.39%.

Kuznetsova et al. [38] used helium as the carrier gas instead of hydrogen, excluding in this way the possibility of water formation. The sample was decomposed at 900°C in the presence of nickelized carbon black. The products in a helium flow passed through a layer of nickelized carbon black kept at 900°C, and then were separated on a column containing 5-Å molecular sieve. The oxygen concentration of the sample is proportional to the area

(height) of the carbon monoxide peak. The error in the determination of oxygen is ±2.2 rel. %. The analysis time is 10 min.

Terent'ev et al. [39] developed a method for the simultaneous determination of nitrogen and oxygen. The sample (5-10 mg)
was decomposed in a platinum or quartz boat under static conditions
in a quartz tube in an atmosphere of oxygen-free helium, under
vacuum (20 mm). At 900°C and in the presence of nickelized carbon
black, the final products from the conversion of oxygen and nitrogen
present in the sample were carbon monoxide and nitrogen. The
authors point out that in some cases (apparently as the result of
the partial hydrogenation of carbon monoxide and dioxide) methane
also appeared among the decomposition products, but in insignificant amounts. After the decomposition of the sample, the oxidation
products were separated chromatographically (carrier gas, helium).
A column filled with 5-Å molecular sieve (0.5-1.0 mm fraction) was
used to separate carbon monoxide and nitrogen. The schematic of
the apparatus is given in Fig. 39 [39].

The calculation of the nitrogen and oxygen concentration of
the sample was carried out from the peak areas by using calibration charts obtained from the combustion of different sample
weights of urea. The area of the nitrogen peak was determined by
multiplying the height of the peak by the width at half height; the
area of the asymmetric carbon monoxide peak was measured with
a planimeter. The accuracy of the determination was checked by
the analysis of known compounds (urea, anthranilic acid, urotro-
pine, hydroquinone, benzoic, oxalic, and salicylic acids). The mean
deviation formed was ±0.24 abs. % and ±0.41 abs. % for nitrogen.

DETERMINATION OF SULFUR

A method for the gas chromatographic determination of sulfur
in organic compounds was developed by Benerman and Meloan [40].
The sample to be analyzed (2-10 mg) was placed in a platinum boat
and burned in oxygen flow (10-12 ml/min) in the presence of a
platinum catalyst at 850°C; the formed sulfur dioxide was determined chromatographically. The oxygen used has to be prepurified
and free of moisture or carbon dioxide.

The products of the primary oxidation of the sample passed
through several layers of platinized asbestos and stars of platinum
wire. Experiments with phenylsulfoxide proved that, using a plati-

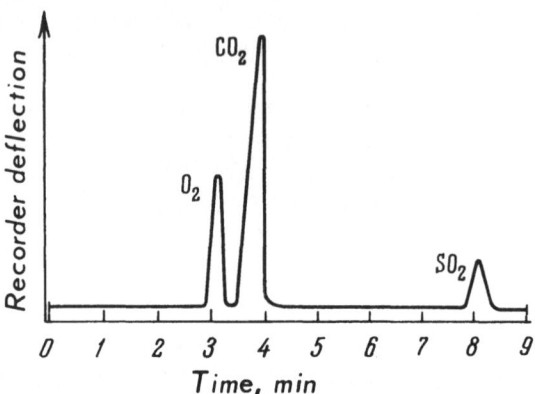

Fig. 40. Chromatogram of the oxidation products in
the simultaneous determination of sulfur and carbon [40].

num catalyst, at 850°C the sulfur content of the sample is quantitatively converted into SO_2, and no SO_3 is formed. If a vanadium
catalyst is used instead of platinum the oxidation temperature must
be increased to 1200°C. Water formed in the combustion is absorbed in a tube containing calcium sulfate in order to prevent the
formation of H_2SO_3. Consequently the formed carbon dioxide and
sulfur dioxide are conducted by the oxygen flow into a U-shaped
trap, cooled with liquid nitrogen, where they condense. The condensed products are introduced into a gas chromatograph and separated at 92°C on a 6-m-long column containing dinonylphthalate as
the liquid phase. A typical chromatogram is shown in Fig. 40 [40].
The sulfur concentration of the sample was determined from the
area of the SO_2 peak, the sample weight, and a calibration factor.

The method was successfully used for the analysis of sulfur-
containing organic compounds, including those which contain fluorine, chlorine, nitrogen, and oxygen. The mean deviation was found
to be 0.37%. The time for analysis is approximately 20 min. The
smallest amount of sulfur which can be determined in a sample is
0.05 mg. The known methods (for example, the lamp method) are
useful for determining smaller concentrations; however, the given
method has good prospects of being developed for the simultaneous
determination of other elements as well as sulfur.

Another method for the determination of sulfur, based on the
preliminary hydrogenation of the sample and determination of the

reaction products by the gas chromatographic method, was developed by Okuno, Morris, and Haines [41]. The method has two variations: determination of the ratio of the sulfur and carbon concentrations (without preliminary weighing of the sample and the determination of the sulfur concentration (absolute) of the analyzed sample.

Both variations consist of the following consecutive steps: the catalytic hydrogenation of the sample resulting in hydrogen sulfide and methane as the principal products, trapping of the hydrogenation products in cool traps, and the chromatographic analysis of these products. Accordingly, the apparatus consists of devices for hydrogenation, trapping, and the gas chromatograph. The individual parts of the apparatus can be made either of metal or glass. Teflon packing was used at all connecting points. Palladium or platinum catalysts on different supports (aluminum oxide, quartz) did not result in quantitative hydrogenations; however, satisfactory results were obtained by using platinum mesh. This catalyst retained its activity for 12 determinations; it can be regenerated by passing air through the heated, catalytic tube. The reactor (45 mm long, 12.5 mm in diameter) was made of quartz. The hydrogenation was carried out in hydrogen flow at 1000°C.

In the first variation of the method (relative analysis), the samples (1 μl) were introduced by injection through a rubber septum. In the second, absolute method, samples weighing 5-10 mg were analyzed. The trapping system consisted of two stainless steel traps. In order to introduce the products into the chromatographic column rapidly, the traps were heated by passing an electrical current through them. The first trap retained hydrogen sulfide; it was filled with stainless steel coils in order to increase the trapping efficiency and cooled with liquid nitrogen. The second trap retained nitrogen and methane; it was filled with 5-Å molecular sieve (0.25-0.5 mm fraction) and cooled with a mixture of dry ice and trichloroethylene. After trapping was completed, the traps were heated rapidly by direct electrical current in order to ensure the introduction of the trapped fraction in the form of sharp zones into the gas chromatograph. The hydrogenation products were analyzed on two columns. The compounds from the first trap were separated at 80°C on a column (180 m long, 6 mm in diameter) containing silica gel. The compounds from the second trap (nitrogen and methane) were analyzed on a 4-Å molecular sieve (0.5-0.6 mm fraction) column. Helium carrier gas and a katharometer detector were in the chromatographic analysis.

It was found that small amounts (less than 1%) of low-boiling hydrocarbons are also formed in the hydrogenation, which are collected in the first trap along with hydrogen sulfide. However, their formation does not disturb the analysis. The amount of light hydrocarbons formed increases when analyzing samples with low sulfur concentration. The silica gel column separates hydrogen sulfide satisfactorily from the other compounds.

The relative sulfur concentration was calculated on the basis of the measured areas of the hydrogen and methane peaks according to the following equation:

$$S = \frac{A_1 \cdot M \cdot 32 \cdot 100}{(A_1 \cdot M \cdot 32) + A_2 \cdot 14},$$

where A_1 and A_2 are the peak areas (heights) for hydrogen sulfide and methane, respectively; M is the calibration factor taking into account the different sensitivity of the katharometer to methane and hydrogen sulfide (this factor has to be determined experimentally by the analysis of a pure compound).

The absolute sulfur concentration was determined by the following equation:

$$S = \frac{A C}{W_s} \cdot 100,$$

where A is the area of the hydrogen sulfide peak, C is the calibration factor, and W_s is the sample weight.

The analytical results for different compounds show that, when determining the relative sulfur concentration, the accuracy of the determination is quite satisfactory and the experimental error does not exceed 10%. Only in one case was a larger error obtained – for thiophene. In the determination of the absolute sulfur concentration the error is ±5%.

This method was used successfully to determine sulfur in petroleum samples. Table 7 lists comparative results of sulfur determinations carried out by the known lamp method and the present method. Both methods gave almost identical results. In the present author's opinion, this method can also be further developed to determine very small amounts of sulfur (in the order of parts per million).

TABLE 7. Quantitative Determination of Sulfur in Petroleum Samples

Petroleum sample	Sulfur concentration, wt. %	
	Lamp method	Hydrogenation followed by gas chromatographic separation
A	0.5	0.46 0.53 0.52
B	0.3	0.27 0.33 0.33
C	0.15	0.15 0.15 0.14
D	0.056	0.056 0.055 0.050

The method for the determination of carbon and sulfur in iron [42] was successfully used for the analysis of samples with carbon concentrations of 0.11–3.28 wt.% and sulfur concentrations of 0.11–0.3 wt.%.

A highly sensitive gas chromatographic method for the determination of carbon (or order ppm) in metallic sodium was developed by Mungall and Johnson [43].

Juranek and Ambrova [44] reported on a gas chromatographic method for the determination of carbon and sulfur in technical iron and its alloys. The sample was burned in oxygen flow serving simultaneously as the carrier gas. The gaseous combustion products (carbon monoxide, carbon dioxide, and sulfur dioxide) were separated chromatographically on a column containing silica gel. The amount of the fractions in the column effluent was recorded with a photocolorimetric cell. This method permits the determinations of carbon concentrations as low as $10^{-6}\%$ by using 1 g steel samples. However, smaller samples can also be used.

Klaas [45] developed a method for the identification and quantitative determination of traces of sulfur-containing compounds in ligroin based on the selective gas chromatographic separation and subsequent oxidation and determination (by titration) of the formed SO_2. Huyten and Rijnders [46] used hydrogenation for the identification of the components after the gas chromatographic separation.

The sulfur-containing compounds were identified by the H_2S peak and the oxygen-containing compounds by the H_2O peak, etc.

It should be mentioned that recently gas chromatographic methods have also been developed for the analysis of other elements (chlorine, bromine) in organic [48] objects as well as inorganic [47] samples.

The large number of studies discussed above indicates the widespread introduction of gas chromatographic methods into the practice of the elemental analysis of pure compounds as well as technical products.

The use of gas chromatographic methods in elemental analysis permits significant shortening of the analysis time, decreases the complexity of the determination, and, in a number of cases (e.g., in nitrogen determination), increases the accuracy of the determination as compared to the classical method. Using the gas chromatographic methods, only very small samples are required, and only the original sample has to be weighed (except when relative values are determined when even this is eliminated). In future studies, particular attention has to be given to the development of methods by which the concentrations of several elements are determined simultaneously and in which the analytical results are given as the ratio of the concentration of the individual elements (H/C, N/C, O/C, etc.).

LITERATURE CITED

1.   A. M. Vogel and J. J. Quattrone, Jr., Anal. Chem., 32:1754 (1960).
2.   R. Levy, Chim. Anal., 46:113 (1964).
3.   Anon., Chem. and Eng. News, 43:33, 41 (1965).
4.   H. J. Francis, Jr., Anal. Chem., 36(7):31A (1964).
5.   A. A. Zhukhovitskii and N. M. Turkel'taub, Gas Chromatography, Gostoptekhizdat, Moscow, 1962.
6.   M. Beroza, J. Gas Chromatog., 2:330 (1964).
7.   V. G. Berezkin, I. A. Musaev, V. S. Tatarinskii, and P. I. Sanin, Gas Chromatography, NIITÉKhIM, Moscow, 1964, p. 25.
8.   A. A. Duswalt and W. W. Brandt, Anal. Chem., 32:272 (1960).
9.   O. E. Sundberg and C. Maresh, Anal. Chem., 32:274 (1960).

10. J. B. Niederl and V. Niederl, Micromethods of Quantitative Organic Analysis, J. Wiley and Sons, New York, 1948.

11. F. M. Power, Ind. Eng. Chem. Anal. Ed., 11:660 (1939).

12. H. S. Haber and K. W. Gardiner, Microchem. J., 6:83 (1962).

13. Y. Mashiko, H. Konosu, and T. Morii, J. Chem. Soc. Japan, Ind. Chem. Sect., 67:555 (1964).

14. H. Konosu and T. Morii, Repts. Govt. Chem. Industr. Res. Inst. Tokyo, 59:351 (1964).

15. B. M. Luskina, S. V. Syavtsillo, A. P. Terent'ev, and N. M. Turkel'taub, Dokl. Akad. Nauk SSSR, 141:869 (1961).

16. M. Vecera, Collect. Czechoslov. Chem. Communs.,26:2278, 2298 (1961).

17. F. Cacace, R. Cipollini, and G. Perez, Science, 132:1263 (1960).

18. F. Cacace, R. Cipollini, and G. Perez, Anal. Chem., 35:1348 (1963).

19. E. P. Skornyakov, Avt. Svid. SSSR 155026 (1962); Byull. Izobr., No. 11 (1963).

20. I. A. Revel'skii, R. I. Borodulina, T. M. Sovakova, and V. G. Klimanova, Dokl. Akad. Nauk SSSR, 159:861 (1964).

21. I. A. Revel'skii, R. I. Borodulina, and T. D. Khokhlova, Neftekhimiya, 4:524 (1964).

22. J. Soucek, Czechoslovak Patent 104920 (1962); RZhKhim. 5G191P, 1964.

23. F. M. Nelsen and S. Groennings, Anal. Chem., 35:660 (1963).

24. D. L. West, Anal. Chem., 36:2194 (1964).

25. J. M. Walker and C. W. Kuo, Anal. Chem., 35:2017 (1963).

26. R. H. Reitsema and N. L. Allphin, Anal. Chem., 33:355 (1961).

27. C. F. Nightingale and J. M. Walker, Anal. Chem., 34:1435 (1962).

28. M. L. Parsons, S. N. Pennington, and J. M. Walker, Anal. Chem., 35:842 (1963).

29. A. A. Moskvina, L. V. Kuznetsova, S. L. Dobychin, and M. I. Rozova, Zh. Anal. Khim., 29:749 (1964).

30. I. E. Pakhomova and M. N. Chumachenko, Izv. Akad. Nauk SSSR, Ser. Khim., 1138 (1965).

31. B. A. Stewart, L. K. Porter, and W. E. Beard, Anal. Chem., 35:1331 (1963).

32. H. Hachenberg and J. Gutberlet, Brennstoff-Chem., 44:235 (1963).

33. M. Simek and K. Tesarik, Collect. Czechoslov. Chem. Communs., 26:1337 (1961).

34.   W. Walisch, Chem. Proc. Eng., 9:828 (1963).
35.   W. Walisch, Chem. Ber., 94:2314 (1961).
36.   J. T. Clerc and W. Simon, Microchem. J., 7:422 (1963).
37.   A. Götz, Z. Anal. Chem., 181:92 (1961).
38.   L. V. Kuznetsova, E. N. Stolyarova, and S. A. Dobychin,
      Zh. Anal. Khim., 20:836 (1965).
39.   A. P. Terent'ev, N. M. Turkel'taub, E. A. Bondarevskyaya,
      and L. A. Domochkina, Dokl. Akad. Nauk SSSR, 148:1316
      (1963).
40.   D. R. Benerman and C. E. Meloan, Anal. Chem., 34:319
      (1962).
41.   I. Okuno, J. C. Morris, and W. E. Haines, Anal. Chem., 34:
      1427 (1962).
42.   W. K. Stuckey and J. M. Walker, Anal. Chem., 35:2015 (1963).
43.   T. G. Mungall and D. E. Johnson, Anal. Chem., 36:70 (1960).
44.   J. Juranek and A. Ambrova, Collect. Czechoslov. Chem. Com-
      mun., 25:2814 (1960).
45.   P. J. Klaas, Anal. Chem., 33:1851 (1961).
46.   F. H. Huyten and G. W. A. Rijnders, Z. Anal. Chem., 205:
      244 (1964).
47.   J. G. Bergman and R. L. Martin, Anal. Chem., 34:911 (1962).
48.   J. C. Mamaril and C. E. Meloan, J. Chromatog., 17:23 (1965).

Chapter VIII

Qualitative Analysis and Change
in Detector Sensitivity

Although the function of a gas chromatographic detector is generally the determination of the concentrations of all separated components leaving the chromatographic column, recently selective detectors (for example, electron capture detectors, spectroscopic detectors, etc.), which detect only a given type of compound, have found wide usage.

A selective detector gives information not only on the amount of the compounds determined but also about their nature, which is of particular importance in the analysis of complex mixtures of unknown composition.

By the use of chemical reactions a nonselective detector can also be converted into a selective detector recording only a certain class of compounds. This is achieved, for example, by installing a reactor in front of the detector in which compounds of a given class are retained selectively (this method was discussed earlier in detail) in which undetectable compounds are converted into detectable compounds (for example, the conversion of carbon monoxide into methane followed by analysis with a flame ionization detector).

The detectors based on the analytical use of chemical reactions (chemical detectors) are usually characterized by a high selectivity.

The utilization of chemical reactions also make possible, in a number of cases, the simplification of the quantitative evaluation of the analytical results. Thus, for example, it is possible, by converting all the separated organic compounds to carbon dioxide and water by combustion and after adsorption of the water, to determine quantitatively such compounds as carbon dioxide. In this case the laborious preliminary determination of the individual response factors of the thermal conductivity detector can be avoided. Here the response factors can be established directly from the concentration of carbon in the molecule of the compound being analyzed.

In practice, qualitative chemical reactions specific for certain functional groups are widely used in gas chromatography in order to determine the nature of the compounds eluted from the column. The use of qualitative chemical reactions can actually be considered as the introduction of an additional selective chemical detector into the chromatographic system.

Thus two basic directions can be distinguished in the use of chemical reactions for detection in gas chromatography: the use of qualitative chemical reactions (selective chemical detection), and the use of chemical reactions for the directed change in the characteristic of the subsequent detection (sensitivity, selectivity).

QUALITATIVE CHEMICAL REACTIONS

AND DETECTORS

Special methods have been developed for the identification of unknown compounds in gas chromatography. The most widespread of these are based on the use of relationships of the type

$$\log V_{i1} = a_k n_i + b_k$$

or

$$\log V_{i1} = a_l \log V_{i2} + b_l,$$

where V_{i1} and V_{i2} are the retention volumes of the compounds on Phase 1 and Phase 2, respectively; n_i is the number of carbon atoms in the given compound. However, it is necessary to know to which homologous series or class of compounds the substance being analyzed belongs in order to be able to select the proper plot, because otherwise the problem is not defined.

TABLE 8. Reagent Characteristics for Functional Group Analysis

Type of compound	Reagent	Characteristics of a positive reaction (color of the solution or precipitate)	Minimum detectable limit, μg	Members of the homologous series tested
Alcohols	$K_2Cr_2O_7-HNO_3$	blue	20	C_1-C_8
	Cerium nitrate	amber	100	C_1-C_8
Aldehydes	2,4-dinitrophenyl-hydrazine	yellow precipitate	20	C_1-C_6
	Schiff reagent	pink	50	C_1-C_6
Ketones	2,4-dinitrophenyl-hydrazine	yellow precipitate	20	C_3-C_8 (methyl ketones)
Esters	ferrous hydroximate	red	40	C_1-C_5 (acetates)
Mercaptans	sodium nitroprusside	red	50	C_1-C_9
	isatin	green	100	C_1-C_9
	lead acetate	yellow precipitate	100	C_1-C_9
Sulfides	sodium nitroprusside	red	50	C_2-C_{12}
Nitriles	iron propyleneglycol hydroxamate	red	40	C_2-C_5
Amines	Grinzberg reagent	orange	100	C_1-C_4
	sodium nitroprusside	1: red	50	C_1-C_4
		2: blue	50	C_1-C_4 (di-methyl and di-amyl amines)
Aromatics	$HCHO-H_2SO_4$	wine-red	20	C_6-C_{10}
Aliphatic unsaturated	$HCHO-H_2SO_4$	wine-red	40	C_2-C_8
Halogen derivatives	alcoholic $AgNO_3$ solution	white precipitate	20	C_1-C_5

In order to determine the type of compound being analyzed (and consequently the selection of the proper plot) Walsh and Merritt [1] used the method of qualitative functional analysis of the fractions isolated after gas chromatographic separation.

The analytical system suggested by Walsh and Merritt is fairly simple. The sample (1 μl) is separated into individual components in the usual chromatographic column and a katharometer is used as the detector. After the katharometer, the gas flow passes through a three-way valve into one of two stainless steel tubes, the ends of which are closed with rubber septums. Up to five injection needles are inserted through each of these stoppers making it possible to split the carrier gas flow and the eluted chromatographic fractions into five, approximately equal, portions. Each needle is immersed in a test tube (3.5 × 1 cm) containing the solution of a group reagent through which the carrier gas bubbles. The results of the qualitative functional analysis of a compound corresponding to a particular peak serves as the basis for selection of the corresponding characteristic plot:

$$\log V_{i1} = a_k n_i + b_k,$$

which is used to identify the given chromatographic peak.

In order to ensure the rapid separation of a fraction zone from the other immediately following it, the gas flow was directed, by means of the three-way valve, into a second tube, while test tubes with fresh reagents were prepared for the first tube-separator.

The characteristics of the reagents used for the qualitative functional analysis are given in Table 8.

As an example the chromatogram of an unseparated, 10-component mixture is given in Fig. 41. The determination of the functional groups showed that Peak I corresponds to alcohol and ketone; Peak II to a ketone and a complex ester; Peak III to a mercaptan; Peak IV to an aldehyde, alcohol, and aromatic hydrocarbon; and, finally, Peak V to an aldehyde and an aromatic hydrocarbon. In using the characteristic plots (Fig. 42) it was established that Peak I corresponds to ethanol and acetone; Peak II to methylpropionate and methylethylketone; Peak III to propylmercaptan; Peak IV to butanol, pentanal, and benzene; and Peak V to hexanal and toluene.

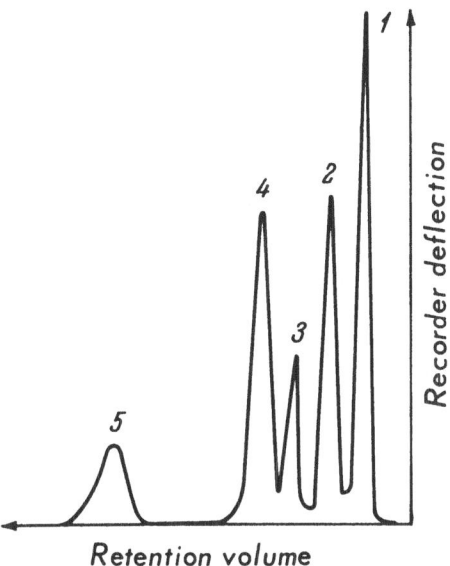

Fig. 41. Chromatogram for a partially separated 10-component mixture [1]. Experimental
conditions: column temperature, 125°C; column length, 2 m; stationary phase, squalane;
sample volume, 1 μl; carrier gas flow rate,
50 ml/min.

In the practical application of this method some difficulties
may arise due to the fact that group reactions are given by more
than one functional group; for example, aromatic and aliphatic unsaturated compounds give the same color reaction. Another problem is to decide whether in case of a positive reaction for two different groups the gas chromatographic fraction corresponds to a
single, bifunctional compound or to two different monofunctional
compounds with identical retention times. Usually these questions
are solved when using the characteristic plots. Thus, for example,
a positive reaction was obtained for a single fraction with the
Schiff and LeRosen reagents. In studying the characteristic curves
for the unsaturated and saturated aldehydes and olefins it was
clearly established that the given chromatographic peak can only
correspond to an unsaturated aldehyde with five carbon atoms since
there are no compounds on the characteristic plots for saturated aldehydes and olefins with the retention time obtained in the analysis.

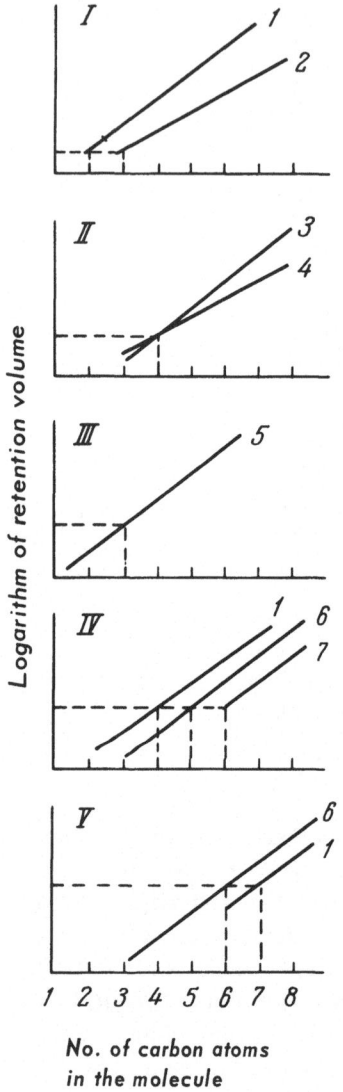

Fig. 42. Characteristic plots for the identification of the compounds corresponding to the peaks in Fig. 41 [1]: 1) alcohols; 2) ketones; 3) carboxylic acid methylesters; 4) methylketones; 5) mercaptans; 6) aldehydes; 7) aromatic hydrocarbons.

The method described above is applicable only for relatively volatile compounds. If high-boiling compounds are to be analyzed, the method must be changed. In this case it becomes necessary to carry out an additional intermediate operation: collection of the chromatographic fraction which is then studied by qualitative reactions at room temperature. Various collection techniques can be used (the use of cooling or an adsorbent).

In using qualitative reactions to study the chromatographic fraction, it is important that the most characteristic reactions are carried out.

Monkman et al. [2] studied various qualitative reactions with the aim of using them for the identification of aliphatic aldehydes in chromatographic fractions. The authors recommend as the most suitable reagents the ethanol solution of 2,4-dinitrophenylhydrazine (minimum detectable limit, 0.08–0.05 μl) and the solution of 0.5 g fuchsin, 9 g sodium bisulfite, and 10 ml concentrated H_2SO_4 in 500 ml water (minimum detectable limit, 0.1 μl). The crystals of the hydrazones formed in the reaction can be studied in ordinary or polarized light under a microscope and their melting points be determined.

The methods for the qualitative analysis of gas chromatographic

fractions are particularly useful in studying complex mixtures containing compounds of different classes. Thus, for example, this
method was used successfully in the analysis of mixtures of different solvents [3] and in combination with thin-layer chromatography for the analysis of the flavor components in orange juice [4].

The identification of the functional groups with the help of
simple, qualitative reactions (the determination of amines, alcohols, carbonyl compounds, esters, and other compounds) were
also described in the monograph of Burchfield and Storrs [5].

The technique of studying the separated chromatographic
fractions by chemical group reagents was refined greatly by
Casu and Cavalotti [6] who suggested a simple automatic apparatus
for the functional groups analysis of chromatographic fractions.
A support layer is wetted with a specific liquid reagent, and this
layer is constantly moving with the speed of the recorder chart
paper before the carrier gas outlet. By comparing the chromatogram and the results of the chemical study, it is easy to determine
the type of compound corresponding to a certain chromatographic
peak. As an example, the chromatogram of a mixture is shown in
Fig. 43 along with the results of the moving layer which was impregnated in this case with a specific reagent for alcohols.

Besides the qualitative functional analysis, the study of
chromatographic fractions for their chlorine, bromine, iodine, sulfur, nitrogen, etc., content is also of interest. A qualitative chemical analysis for these elements gives valuable information on the
qualitative composition of the sample.

Berezkin, Janák, and Hřivnač [7] suggested a single method
for the qualitative identification of compounds containing halogens,
sulfur, and nitrogen. After the usual chromatographic separation
and detection by a katharometer, the carrier gas (nitrogen) flow is
united with oxygen (1:1), and enters a reactor containing a platinum
catalyst (wire). The combustion products bubble through a dilute
alkaline solution, which consequently is studied with the help of
qualitative reactions for the presence of anions (e.g., chlorine,
nitrite) during the destruction of the molecules of the corresponding heterocompounds.

If the concentration of the anion is determined quantitatively,
with a photometric method, this result might also permit us to

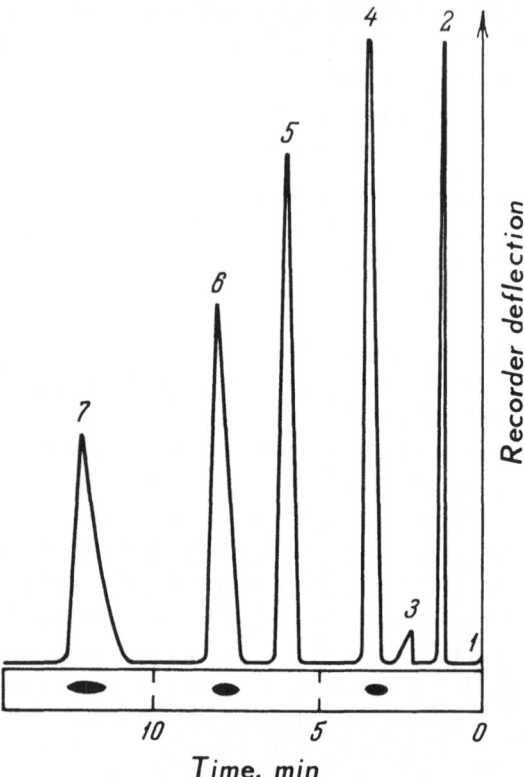

Fig. 43. Chromatogram and results of the qualitative
analysis by separated gas chromatographic fractions
on a moving plate for the hydroxyl group [6]: 1) air;
2) ether; 3) water; 4) ethanol; 5) methylethylketone;
6) n-propanol; 7) butanol-2.

calculate the actual amount of the substance present which contains
the heteroatom.

Such quantitative determination can also be utilized in the
standard variation of gas chromatography. Thus, for example,
Parker et al. [8] used the methods of quantitative spectrophotom-
etry for the analysis of the alkyl lead compounds in gasoline from
their chromatographic fraction. The gasoline sample (50-100 μl)
was separated at 70°C on a chromatographic column containing 40%
Nujol on Chromosorb support. The tetraalkyl lead fractions were
collected in separate test tubes and mixed with a 0.02 N solution of
iodine in methyl alcohol (5 ml). After the reaction was completed

the products were analyzed photometrically. In the analysis of a mixture of tetramethyl lead and tetraethyl lead in gasoline, the concentration of the latter was determined from the difference between the total lead concentration in the gasoline and the lead concentration corresponding to the tetramethyl lead, since under the chromatographic condition used, the retention time of tetraethyl lead is too long.

The functional group analysis of the compounds after gas chromatographic separation can also be carried out by thin-layer or paper chromatography (e.g., Janák [9]), which can also be considered as methods of qualitative analysis.

The technique developed by Janák [10, 11] for the separation of complex mixtures – the so-called multiple chromatographic method – deserves special attention. In multiple chromatography, the mixture is first separated on a column by the methods of gas-liquid chromatography. The fractions leaving the chromatographic column are continuously deposited on a moving plate with an adsorbent or paper. The compounds, which are deposited in this way on the starting line, are then separated by the usual methods of paper or thin-layer chromatography. The chromatogram obtained after development is two-dimensional; in one direction the separation is carried out according to the molecular weight of the compound (gas-liquid chromatography on a nonpolar phase), and in the other according to the type of functional group (thin-layer chromatography). The separated substances on the thin-layer chromatogram can be observed and identified colorimetrically, by crystallographic methods, or by separation from the layer for further gas-chromatographic studies.

The separation of the compounds in thin-layer chromatography depends mainly on the structure and type of the functional substituting groups, while the molecular weight of the components is usually not a determining factor. The polarity of the adsorbent in thin-layer chromatography is often much greater than the polarity of any of the known phases in gas-liquid chromatography. Moreover, in gas-liquid chromatography a nonpolar phase separates the sample components in the first approximation according to the number of carbon atoms in their molecules, i.e., according to their molecular weight. Therefore, the combination of both methods increases the analytical possibilities of the new method, particularly for the identification of the components of complex mixtures.

In the work of Janák [10], the initial separation was carried out on a stainless steel (100×0.6 cm) column fitted with 10% of a methylphenyl silicone polymer (mol. wt:370,000) on Celite (0.2–0.3 mm fraction). Argon or nitrogen was used as the carrier gas.

The outlet of the chromatographic column was placed 2 mm above the surface of a silica gel layer; the distance from the side of the plate was 10–15 mm. For a carrier gas flow rate of less than 1 ml/sec no destruction of the silica gel layer was observed. Under these conditions the width of the initial zone was less than 3 mm. The silica gel plates were placed on a moving belt. The fractions from the chromatographic column were continuously deposited on the silica gel layer as the belt with the plate moved along. The velocity of the plate could be varied between 2 and 75 mm/sec.

The subsequent separation was carried out by the usual methods of thin-layer chromatography. Powdered RNN silica gel (Czechoslovakia) was deposited on a plate of mirror glass of dimensions 100×200 mm or 200×200 mm. The particle size of the silica gel was 0.5–0.15 mm, and the thickness of the layer was 0.6–0.9 mm. In order to decrease the transverse diffusion, metallic, nickel-plated plates with grooves were used in some cases. The width of the grooves was 3 mm; the depth was 1 mm; the distance between them was 1 mm. The plates were placed in a $150 \times 300 \times 150$ mm gas chamber. The slope of the plate was 20–30°C. n-Hexane, cyclohexane, benzene, chloroform, acetone, ether, or a benzene:cyclohexane mixture (1:1) were used as solvents. The time of the chromatographic separation on the plate was 8–15 min.

The separated zones on the plates were observed by forming colored complexes with tetracyano-ethylene (TCE). A solution of TCE in benzene (3.6 g in 100 ml) was used for development. The solution was dropped on the places where the appearance of colored spots was expected. After evaporation of the benzene in a drying chamber, the colored spots of the complexes appeared. The intensity of the color remained constant for 1–2 hr for all compounds except phenols, for which the color became deeper.

The suggested method of multiple chromatography was successfully used to study the composition of coal tar. The further development of this method is discussed by Janák et al. [11].

It should be pointed out that in a number of cases it is convenient to use thin-layer chromatography, first followed by gas-liquid chromatography. Thus, for example, Mangold and Kammereck [12] in the analysis of mixtures of saturated and unsaturated aliphatic acids, first separated the methylesters into groups which contain the same degree of unsaturation. Consequently, these groups of acids were separated by gas-liquid chromatography.

We should also mention here those selective chemical detectors which are based on a given chemical reaction and which permit the direct, quantitative as well as qualitative evaluation of the compounds being eluted. Detectors of this type can be either differential (e.g., a thermochemical detector based on the catalytic combustion of organic compounds on the surface of a platinum catalyst) or integral (e.g., a titration detector). Obviously the flame ionization detector may also be considered as a selective physico-chemical detector for the determination of organic compounds.

For the analysis of volatile aliphatic acids James and Martin [13] used an automatic titration cell. The compounds eluted from the column entered a chamber containing either an aqueous or a nonaqueous solvent. The color indicator of the pH of the medium, in conjunction with a photoelement and a relay, controlled the addition of the titrating solution. The position of the plunger of the buret, made in the form of a syringe, was recorded. Such a titration detector records the integral curve corresponding to acids eluted from the column. It permits the determination of the acids (or amines) selectively from their mixture with other compounds. The operating temperature of the cell is limited by the vapor pressure of the titrating medium. The sensitivity of the detector is 0.002-0.02 mg acid or alkali. The use of a coulometric titration detector is described by Thielemann [14]. The method of recording the chromatographically separated methylchlorosilanes by a change in the conductivity of the solution, which is the result of the formation of hydrochloric acid from the hydrolysis of the chlorosilanes, was suggested by Garzo et al. [15].

Kelker [16] suggested using the Beilstein test for the qualitative characterization of a halogen-containing sample. Hydrogen was used as the carrier gas; its flow from the outlet of the katharometer entered a flame jet (a syringe needle, 8 cm long). The

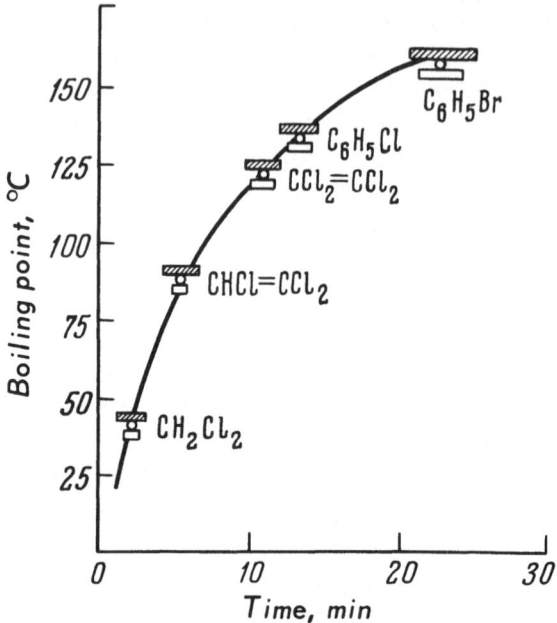

Fig. 44. Correlation of the retention time of halogen-
containing compounds and the boiling point [18]: O,
retention time; ▨, width recorded by the katharometer;
□, peak width corresponding to the duration of the green
color of the flame (Beilstein flame detector).

presence of small amounts of chlorine-containing components can
be established from the green color of the flame (the copper coil
used in the Beilstein test was heated to a high temperature by an
additional jet). This method has great significance when it is necessary to determine the presence of halogen derivatives in a complex mixture of 20-30 components. This method was developed in
greater detail by Chovin et al. [17] and by Gunther et al. [18].

In the system described by Gunther et al. [18] the column
effluent or a part of it entered a tube ending with a copper mesh
which was attached above a burner. Upon elution of a halogen-containing compound, the flame became greenish in color. The range
of optimum detection was 0.17-50 μg/sec of bound chlorine. Organic compounds containing CN, SCN, as well as groups which
form cyanides in a flame, result in an analogous color.

The data presented in Fig. 44 [18] characterize the definiteness of the identification of halogen-containing compounds. The sensitivity of the determination is equal to or better than the sensitivity of a katharometer. A highly sensitive detector based on the Beilstein test was described by Huyten and Rijnders [19]. This detector can determine halogen-containing compounds with a minimum concentration of 5-80 parts per million, and can also be used in conjunction with capillary columns.

Apparently the chemical reaction of the combustion products from halogen- and phosphorus-containing organic compounds with sodium hydroxide deposited on a platinum collector electrode increases the sensitivity of a flame ionization detector to compounds of this type [20]. Such systems can also be used for the qualitative determination of these compounds.

USE OF CHEMICAL REACTIONS
IN FRONT OF THE DETECTOR

If the compounds eluted from the column are converted into new compounds in a reactor, then the usual sensitivity of the detection changes, sometimes very sharply.

The detector signal for the conversion of the compounds is proportional to $m_i P_0$, where m_i is the number of moles of the new compound formed from one mole of the initial compound and P_0 is the molar sensitivity of the detector to the new compound.

The use of converters of this type has three purposes: first, to change the sensitivity of the detector (conversion of the undetectable compounds into detectable compounds and vice versa); second, to improve the measurement characteristics of the detector; and third, to expand the field of application of the detector and to simplify its construction.

In connection with the fact that in the practice of gas chromatography the katharometer and the flame ionization detector are the two most widely used detectors, these converters will be studied in detail more applicable to these two detectors.

The katharometer is a simple and reliable detector for the majority of compounds which are present in a mixture if their amount is not too small. The main shortcoming of the katharometer is the necessity of utilizing individual correction factors in the calculations (e.g., [21, 22]) which depend on the type of katharometer and the experimental conditions. This shortcoming can be avoided if all sample components are converted into a certain single compound. When working with a katharometer, conversion of organic substances to carbon dioxide is used most frequently. Sometimes conversion to hydrogen or methane is also utilized.

As a result of the conversion, the necessity of prolonged and laborious calibration of the apparatus is eliminated, because now, each peak corresponds to one single substance and the concentration of the individual components in weight percent can be obtained directly from the peak areas. Besides this, the sensitivity of the detection is increased both because of the increase in the concentration of the carbon dioxide being measured (one molecule of an organic compound usually gives several molecules of carbon dioxide upon combustion) and because of the selection of more optimum conditions for the measurement (low detector temperature, high filament current, etc.). Finally, by using this method, the construction of the katharometer can be simplified because the conversion of each (even high-boiling) fraction to carbon dioxide permits the katharometer to be thermostatted, for example, at room temperature, while the chromatographic column is kept at a much higher temperature. When it is necessary to make an additional study of the compounds being analyzed (for example, by means of qualitative reactions), it is possible to split the column effluent and to convert only part of it.

Martin and Smart [23] were the first to utilize the combustion of organic compounds to carbon dioxide in front of the detector. Simmons et al. [24] recommended that a microanalytical furnace be used for a quantitative combustion. The combustion tube is filled with 7 g copper oxide wire impregnated with a solution of ferric nitrate calculated to give a 1% concentration of iron on the copper oxide. The combustion temperature is 725-825°C. The volume of a single sample was approximately $3\mu l$. The combustion on copper oxide in helium flow is quantitative up to an accumulated total sample volume of 0.25-0.30 ml. Consequently, the copper oxide has to be regenerated in air. After 10-15 cycles, the copper-oxide packing begins to sinter and should be replaced. The effluent from the converter containing the combustion products (carbon dioxide and water) passes through a tube (15×0.6 cm) filled with a drier (e.g., magnesium perchlorate) and enters the sensing cell of the katharometer.

Simmons et al. [24] placed an adsorber with Ascarite (volume, 15 ml) between the sensing and reference cells of the katharometer to absorb the carbon dioxide. The advantage of this system is that incomplete combustion (the formation of carbon monoxide which is not adsorbed on Ascarite) will result in the appearance of a negative peak on the chromatogram indicating that either the copper oxide should be regenerated or that the sample size is too large. The instability of the zero line is usually related to the adsorbers. The conversion of the fractions in the column effluent to carbon dioxide in front of the detector was used successfully in a number of studies (e.g., [25-27]).

In a number of papers [28, 29] quantitative oxidation was used to carry out the continuous, radiometric detection of compounds labeled with ^{14}C or ^{3}H. The combustion to carbon dioxide permits the use of a simple flow counter to measure radioactivity.

Green [30] suggested carrying out a double conversion of the separated fractions to increase the sensitivity of the detection. After oxidizing the fraction of the organic compounds to carbon dioxide and water, the gas flow enters the second section of the reactor containing powdered, reduced iron, where water is reduced to hydrogen. The reactor temperature is 700-800°C. After the reactor, the gas flow passes through a short column filled with soda lime which removes carbon dioxide. From here, the effluent is conducted to the katharometer detector where the hydrogen peaks are recorded. Nitrogen is used as the carrier gas.

Chromatograms obtained by using different methods of conversion are shown in Fig. 45. This figure shows that the application of conversion methods increases the sensitivity of detection, thus making the use of smaller samples possible. This fact increases the efficiency of the chromatographic separation.

Franc and Wurst [31] introduced a silver layer for the analysis of chlorine- and silicon-containing compounds and suggested the following packing for the combustion tube: copper-oxide wire (10 cm), silver on iron (5-cm layer; temperature of the silver layer, 480°C), iron filings.

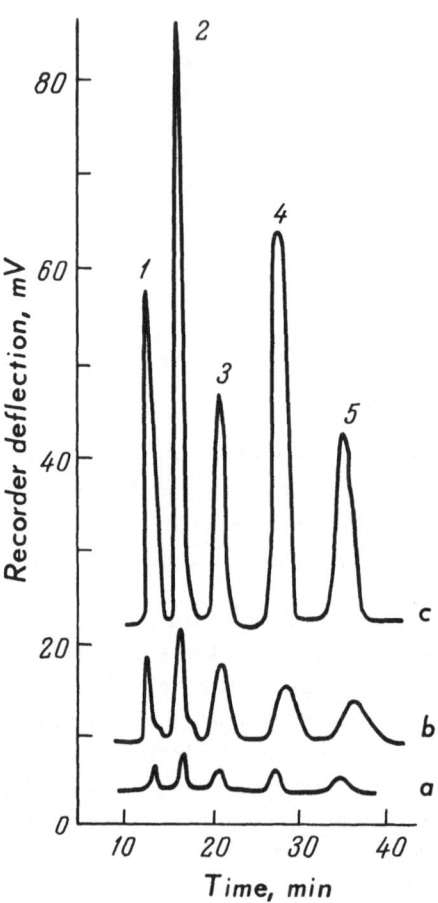

Fig. 45. Change in the sensitivity of determination of organic compounds by using different conversion methods [30]:   a) without
conversion; b) conversion to carbon dioxide;
c) conversion to hydrogen. 1) Cinnamalde-
hyde; 2) butylbenzoate; 3) coumarin; 4)
phenylpropylbenzoate; 5) naphthylphenyl-
ketone. Sample volume, 1 μl; column material, 20% silicone oil on celite-545; carrier gas, nitrogen (13 ml/min); katharometer
detector.

It should be pointed out that in spite of the quantitative results obtained with the double conversion of organic compounds to H_2 [31, 32] complications are possible which are associated with the fact that the equilibrium constant $K = [H_2]/[H_2O]$ for the reaction $Fe + H_2O \rightleftharpoons FeO + H_2$ is approximately equal to 2, and the equilibrium constant $K = [CO]/[CO_2]$ for the reaction $Fe + CO_2 \rightleftharpoons FeO + CO$ is equal to 1.5 [33]; i.e., the reaction does not go to completion. However, since the thermal conductivity of CO is much less than that of H_2, the presence of CO in the H_2 fraction can be ignored. Besides this, quantitative detection with respect to hydrogen can also be obtained when the reaction goes to equilibrium because the degree of conversion will still be the same for all of the compounds. Furthermore, it is also possible that the reaction conversion for $H_2O \rightarrow H_2$ takes place under chromatographic conditions, and therefore goes to completion.

Ridgway and Zlatkis [34] suggested using destructive demethylation of hydrocarbons and other organic compounds in front of the katharometer. According to them, the compounds are converted to CH_4 in a special reactor at 230-420°C on a nickel catalyst in H_2 flow. The flow rate of the H_2 must be sufficiently low so that the complete conversion of the sample components is ensured. This method was used successfully for the analysis of mixtures of $C_1 - C_{12}$ paraffins and olefins, $C_5 - C_8$ naphthenes, aldehydes, alcohols, and ketones. It was shown, using hexane isomers as an example, that as the result of the conversion, the sensitivity of detection increases (the height of the peaks on the chromatogram are doubled or tripled). Conversion to methane was also used by Zlatkis et al. [35, 36] for the analysis of amino acids and the volatile aldehydes formed in the ninhydrin method of amino acid analysis.

Methods for the conversion of carbon monoxide and dioxide [37, 38] and of carbon disulfide and oxysulfide [39] into methane on a nickel catalyst in hydrogen flow are described in the literature. Methods have also been suggested for the determination of traces of hydrogen by its catalytic conversion into methane in a carbon monoxide atmosphere [37] and for the analysis of hydrogen cyanide and dicyanogen after the hydrogenation of these compounds to methane [39].

Besides the widely used hydrogenation reaction for the conversion of undetectible compounds into organic compounds, which

can be determined with a flame ionization detector, the reaction of
water with calcium carbide, resulting in the formation of acetylene
[40], and the double-conversion reaction permitting the recording
of oxygen with a flame ionization detector [41] are also reported.
The method of Gudzinowicz and Smith [42] in which nonradioactive
compounds are analyzed by a radiometric method also belongs in
this category. Their method is based on the fact that an inorganic
oxidizing agent decomposes the clathrate of radioactive krypton,
resulting in the emission of a radioactive isotope, which is then
recorded with a radiometric counter.

The examples discussed above indicate the great practical
value of using methods of reaction gas chromatography in the field
of detection. The main direction of further development in this
field will obviously be the development of systematic qualitative
and quantitative methods and techniques to carry out all the operations for functional analysis of the eluates (particularly a microanalytical variation) and the development of new conversion methods for the analysis of inorganic compounds with highly sensitive
ionization detectors.

LITERATURE CITED

1.   J. T. Walsh and C. Merritt, Jr., Anal. Chem., 32:1378 (1960).
2.   J. L. Monkman, L. Dubois, and T. Techman, Chem. Canada,
     13(5):92 (1961).
3.   J. K. Haken and T. R. McKay, J. Oil & Colour Chem. Assoc.,
     47:517 (1964).
4.   J. A. Attaway and R. W. Wolford, in "Gas Chromatography
     1964," ed. A. Goldup, Institute of Petroleum, London, 1965,
     p. 170.
5.   H. P. Burchfield and E. Storrs, Biochemical Applications of
     Gas Chromatography, Academic Press, New York, 1962
     [Russian translation: Izd. "Mir," Moscow, 1964].
6.   B. Casu and L. Cavalotti, Anal. Chem., 34:1514 (1962).
7.   W. Berezkin, J. Janák, and M. Hřivnač, XIX Celostatni
     chemicky sjerd, 1962.
8.   W. W. Parker, G. Z. Smith, and R. L. Hudson, Anal. Chem.,
     33:1170 (1961).
9.   J. Janák, Nature, 195:696 (1962).
10. J. Janák, J. Chromatog., 15:15 (1964).

11. J. Janák, I. Klimes, and K. Hana, J. Chromatog., 18:270 (1965).

12. H. K. Mangold and R. Kammereck, Chem. & Ind. (London), 1032 (1961).

13. A. T. James and A. J. P. Martin, Biochem. J., 50:679 (1952).

14. H. Thielemann, Chem. Techn. (Berlin), 14:162 (1962).

15. T. Garzo, F. Till, and J. Till, Magyar Kémiai Folyóirat, 68:327 (1962).

16. H. Kelker, Angew. Chem., 71:218 (1959).

17. P. Chovin, J. Lebbe, and H. Moureau, J. Chromatog., 6:363 (1961).

18. F. A. Gunther, R. C. Blinn, and D. E. Ott, Anal. Chem., 34: 302 (1962).

19. F. H. Huyten and G. W. A. Rijnders, Z. Anal. Chem., 205: 244 (1965).

20. A. Karmen and L. Giuffrida, Nature, 201:1204 (1964).

21. G. R. Jamieson, J. Chromatog., 3:464 (1960).

22. A. E. Messner, D. M. Rosie, and P. A. Argabright, Anal. Chem., 31:230 (1959).

23. A. E. Martin and J. Smart, Nature, 175:422 (1955).

24. M. C. Simmons, L. M. Taylor, and N. Nager, Anal. Chem., 32:731 (1960).

25. I. R. Hunter, V. H. Ortegren, and J. W. Pence, Anal. Chem., 32:682 (1960).

26. W. Stuve, in "Gas Chromatography 1958," ed. D. H. Desty, Butterworths, London, 1958, p. 178 [Russian translation: IL, Moscow, 1961].

27. S. I. Krichmar and M. I. Beilina, Zavodsk. Lab., 26:1171 (1960).

28. F. Cacace, R. Cipollini, and G. Perez, Science, 131:732 (1960).

29. F. Drawert and O. Bachmann, Angew. Chem., 75:717 (1963).

30. G. E. Green, Nature, 175:295 (1957).

31. J. Franc and W. Wurst, in "Gas Chromatography," Tr. I Vses. Konf., Izd. Akad. Nauk SSSR, Moscow, 1960, p. 289.

32. V. G. Berezkin and L. S. Polak, Tr. Kom. Anal. Khim. Akad. Nauk SSSR, 13:205 (1963).

33. M. Kh. Karapetyants, Chemical Thermodynamics, Goskhim-izdat, Moscow, Leningrad, 1958.

34. J. A. Ridgway and A. Zlatkis, U. S. Pat. 3030191 (1962).

35.   A. Zlatkis and J. A. Ridgway, Nature, 182:130 (1958).

36.   A. Zlatkis, J. F. Oro, and A. P. Kimball, Anal. Chem., 32: 162 (1960).

37.   U. Schwenk, H. Hachenberg, and M. Förderreuther, Brenn-stoff-Chem., 42:295 (1962).

38.   K. Porter and D. H.Volman, Anal. Chem., 34:748 (1962).

39.   K. Tesarik, in "Gas Chromatographie 1965," ed. H. G. Struppe and D. Obst, Akademie Verlag, Berlin, 1965, Supplement, p. 89.

40.   H. S. Knight and F. T. Weiss, Anal. Chem., 34:749 (1962).

41.   V. G. Berezkin, A. E. Mysak, and L. S. Polak, Izv. Akad. Nauk SSSR, Ser. Khim., No. 10:1871 (1964).

42.   B. J. Gudzinowicz and W. R. Smith, Anal. Chem., 35:465 (1963).

Chapter IX

Selective, Chemically Active Sorbents in Gas Chromatography

The actual separation of the components in gas chromatography is determined by the ratio of two opposed factors: the difference in the rates of movement of the chromatographic zone along the column (selectivity of the stationary phase), and their progressive diffusion (efficiency of the column).

The degree of separation Δ of two compounds which are present in equal quantities in the sample can be evaluated by the equation (see, for example, [1]):

$$\Delta = 1 - 2\exp\left[-2N\left(\frac{r-1}{r+1}\right)^2\right] \qquad (1)$$

where N is the number of theoretical plates and $r \approx t_2/t_1$. The selectivity of the stationary phase is characterized by the value of r. The ratio

$$\frac{\partial\Delta}{\partial r} : \frac{\partial\Delta}{\partial N} = 4N(r^2 - 1) \gg 1 \qquad (2)$$

characterizes the role of both of the above factors. From Eq. (2) it follows that the actual separation is determined to a greater degree by the selectivity of the stationary phase.

A sharp increase in the selectivity of the phase can be achieved by utilizing chemical as well as physical factors.

The selectivity for the separation of compounds with similar physical properties increases greatly by using specific, chemically active stationary phases which form unstable chemical compounds with the sample component during the separation process. Among phases of this type, complex forming substances are most widely used in gas chromatography.

Bradford et al., [2] were the first to use silver salts in a polar solvent for the separation of hydrocarbon gases. The silver salts form complexes with the unsaturated compounds [3] permitting the selective separation of the latter (for example, α- and iso-butylenes).

The partition coefficient, when using the solution of a complexing agent in a nonvolatile solvent as the stationary phase, is determined by the solution of the sample in this solvent and by the formation of the complex. The relationship of the distribution coefficient D of the dissolved substance A to its distribution coefficient in a pure, inert solvent D_0 and the stability constant of the complex K is expressed by the equation (see, for example, [4]):

$$D = D_0 + D_0 K^{X_B} \tag{3}$$

where X_B is the number of moles of the complexing agent B in the stationary liquid phase.

In studying the stability constant for the complexes of a number of unsaturated $C_4 - C_5$ hydrocarbons with silver nitrate in ethyleneglycol and diethyleneglycol. Genkin and Boguslovskaya [4] found that the constants are independent of the solvent and the undissociated molecules of the salt as well as individual silver ions participate in the complex formation.

The use of solutions of silver salts in different polar solvents (diethyleneglycol, benzylcyanide, triethyleneglycol, etc.) for the separation of unsaturated compounds was studied in a number of papers [5-7]. The detailed study of Smith and Ohlsen [7], who obtained data on the retention values for 75 unsaturated compounds in the $C_2 - C_7$ range should be pointed out in particular. The high selectivity of the silver phases for the separation of deuterated isomers of the unsaturated compounds was shown by Cvetanovic et al. [8]. Table 9 lists the relative retention values of deuterated olefins with respect to their light isomers. In all cases, a com-

TABLE 9. Relative Retention Values of Deuterated Olefin Isomers with Respect to the Corresponding Light Isomers [8]

Compound	Temperature	
	16 ± 3°C	1 ± 1°C
1,2-Dideuteroethylene	1.07	1.07
Tetradeuteroethylene	1.15	1.14
Hexadeuteropropylene	1.12	1.11
Octadeuterobutene-2, trans	1.11	1.09
Octadeuterobutene-2, cis	1.10	1.08
2-Methyl-octadeuterobutene-2	1.10	1.07
3-Methyl-octadeuterobutene-2	1.11	1.10
Octadeuteropentene-2, trans	1.10	1.08
Octadeuteropentene-2, cis	1.08	1.07

pletely satisfactory separation was obtained. Obviously it is possible to achieve a finer separation by using a highly effective capillary column.

Stationary phases containing silver salts are widely used in chromatographic systems incorporating composite columns [9-11].

The shortcomings of phases of this type are their low temperature maximums (60-80°C) and the irreversible reactions with acetylene derivatives.

Another interesting example of the use of complexes in gas chromatography was the direct use of melted heavy metal salts [12]

TABLE 10. Relative Retention Values of Some Amines and Alcohols on Apiezon-L (I) and Manganese Stearate (II) at 156°C. Standard: Mesitylene [12]

Compound	Boiling point, °C	Stationary phase	
		I	II
β-Picoline	143	0.535	7.46
p-Picoline	143	0.540	9.67
2,6-Lutidine	143	0.571	0.773
n-Butyl alcohol	118	0.123	0.356
Isobutyl alcohol	108	0.111	0.244

and complexes of metals with a variable valence [13] as the stationary phases. These stationary phases can be used at high temperatures. Relative retention values on manganese stearate and Apiezon-L are given in Table 10 to characterize their selectivity.
These data show that the use of manganese stearate ensures the
selective separation of amines and alcohols.

In studying a number of complexes (n-dodecylsalicyaldimines
of nickel, palladium, platinum, and copper, as well as methyl-n-
octylglyoximines of nickel, palladium, and platinum) as stationary
phases, specific retention values of amines, ketones, alcohols, and
aldehydes, were determined by Cartoni et al. [13].

It is noted that the improvement in the separation of some
compounds, for example of amines, can also be achieved by using
a support impregnated with copper salts (which form complexes of
different stability) prior to the application of the liquid phase [14].

It is also expedient to use chemically active phases for the
separation of inorganic compounds. This was first shown by
Glueckauf and Kitt [15] using, as an example, the separation of hydrogen isotopes [15] on a column containing platinum black.

Zhukhovitskii et al. [16] used stabilized bull's blood, the
hemoglobin of which forms an unstable complex compound with
oxygen, to separate oxygen and argon, which cannot be separated
under ordinary conditions on molecular sieves.

Takashima et al. observed the chromatographic separation of
nitrogen and oxygen on manganese oxide at elevated temperatures [17].
Zhukhovitskii et al. [18] investigated the possibility of using the
phenomenon of activated adsorption in gas chromatography. In
studying the dependence of the retention volume of argon and oxygen
on hopcalite on the temperature, it was shown that in the range of
200–320°C an anomalous increase in the retention volume of the
oxygen was observed, which indicated that an activated chemical
process took place.

In conclusion it must be pointed out that the use of complex
formation in gas chromatography is presently limited; the broad
development of this direction is just beginning. The number of investigated complexing agents is very small, and the overall limits
of the method, the optimum conditions for the analysis, etc., are

not yet determined. However, the results obtained leave no doubt
as to the rapid development of this field in the near future.

LITERATURE CITED

1. H. G. Struppe, Abhandl. Dtsch. Akad. Wiss. Berlin Kl. Chem.
    Geol. und Biol., (9):28 (1959).
2. B. W. Bradford, D. Harvey, and D. E. Chalkley, F. Inst.
    Petrol., 41:90 (1955).
3. L. Endryus and R. Kifer, Molecular Complexes in Organic
    Chemistry, Izd. "Mir," Moscow, 1967.
4. A. N. Genkin and B. I. Boguslovskaya, Neftekhimiya, 5:897
    (1965).
5. M. E. Bednas and D. S. Russel, Can. J. Chem., 36:1272 (1958).
6. E. Bendel, M. Kern, R. Janssen, and G. Steffan, Angew. Chem.,
    74:905 (1962).
7. B. Smith and R. Ohlsen, Acta Chem. Scand., 16:351 (1962).
8. R. J. Cvetanovic, F. J. Duncan, and W. E. Falconer, Can.
    J. Chem., 41:2095 (1963).
9. J. A. Barnard and H. W. D. Hughes, Nature, 183:250 (1959).
10. V. G. Berezkin and L. S. Polak, Khim. i Tekhnol. Topliv i
    Masel, No. 4:14 (1961).
11. M. S. Vigdergauz and K. A. Gol'bert, Khim. i Tekhnol.
    Topliv i Masel, No. 11:67 (1961).
12. D. W. Barber, C. S. G. Phillips, G. F. Tusa, and A. Verdin,
    J. Chem. Soc., 18 (1959.
13. P. Cartoni, S. R. Lowrie, C. S. G. Phillips, and L. M. Venanzi,
    in "Gas Chromatography 1960," ed. R. P. W. Scott, Butter-
    worths, London, 1960, p. 273 [Russian translation: Izd.
    "Mir," Moscow, 1964, p. 362].
14. B. G. Belen'kii, A. G. Vitenberg, L. D. Turkova, and
    N. N. Chernyshkov, Izv. Akad. Nauk SSSR, Ser. Khim., No.
    2:269 (1967).
15. E. Glueckauf and G. P. Kitt, in "Vapour Phase Chromatog-
    raphy," ed. D. H. Desty, Butterworths, London, 1957, p. 422.
16. A. A. Zhukhovitskii, N. M. Turkel'taub, and A. F. Shlyakhov,
    Khim. i Tekhnol. Topliv i Masel, No. 6:7 (1962).
17. I. Takashima, A. Koga, and J. Kaneko, J. Chem. Soc. Japan,
    65:1223 (1962).
18. N. T. Ivanova, A. A. Zhukhovatskii, and S. V. Syavtsillo,
    Zavodsk. Lab., 32:136 (1966).

Index